ION IMPLANTATION IN MICROELECTRONICS

A COMPREHENSIVE BIBLIOGRAPHY

COMPUTER SCIENCE INFORMATION GUIDES

Volume 1—ION IMPLANTATION IN MICROELECTRONICS
A Comprehensive Bibliography
A. H. Agajanian

ION IMPLANTATION IN MICROELECTRONICS

A COMPREHENSIVE BIBLIOGRAPHY

A. H. Agajanian

IBM General Technology Division
Hopewell Junction, New York

IFI/PLENUM • NEW YORK-WASHINGTON-LONDON

Library of Congress Cataloging in Publication Data

Agajanian, A. H.
Ion implantation in microelectronics.

(Computer science information guides ; v. 1)
Includes indexes.
1. Semiconductor doping—Bibliography. 2. Ion implantation—Bibliography. I. Title. II. Series.
Z5838.S4A38 [TK7871.85] 016.621381'71 81-10753
ISBN 978-1-4684-6134-3 AACR2

ISBN 978-1-4684-6134-3 ISBN 978-1-4684-6132-9 (eBook)
DOI 10.1007/978-1-4684-6132-9

A Division of Plenum Publishing Corporation
227 West 17th Street, New York, N.Y. 10011

To
Robert B. Murphy

PREFACE

During the past ten years the use of ion implantation for doping semiconductors has become an active area of research and new device development. This doping technique has recently reached a level of maturity such that it is an integral step in the manufacturing of discrete semiconductor devices and integrated circuits. Ion implantation has significant advantages over diffusion such as: precision, purity, versatility, and automation; all of which are important for VLSI purposes. Ion implantation has also found new applications in magnetic bubble domain materials, superconductors, and materials synthesis.

This book is a comprehensive bibliography of 2467 references of the world's literature on ion implantation as applied to microelectronics. This compilation will easily enable researchers to compare their work with that of others. For easy access to the needed references, the contents are divided into fifty-two subject headings. The main categories are: bibliographies, books and symposia, review articles, theory, materials, device applications, and equipment. An author index and a subject index are also given to provide easy access to the references.

The literature from January 1976 to December 1980 is covered. The literature prior to 1976 is the subject, in part, of a previous book by the author (1). The main sources searched were: Physics Abstracts (PA), Electrical and Electronics Abstracts (EEA), Chemical Abstracts (CA), Nuclear Science Abstracts (NSA), and Engineering Index. The volumes and numbers of the abstracts are given to provide access to the abstracts.

The author would like to thank Mr. R. B. Murphy, IBM East Fishkill Library manager, for encouragement and support, my wife Else for encouragement and assistance, my colleagues Dr. W. K. Chu, Dr. W. W. Hicks, and Dr. B. J. Masters for helpful discussions. I would also like to thank Mr. L. A. Plaushin for editorial assistance, and the IBM East Fishkill Library staff for their valuable assistance.

AVAILABILITY OF DOCUMENTS

The references in this book consist of technical articles presented in journals, books, symposia proceedings, technical meetings, reports, patents and theses. If a reader does not have access to a library to obtain copies of the references, we suggest the following sources:

1. Journal articles, books, symposia proceedings:

   Library of Congress
   Photoduplication Service
   Washington, D. C. 20540

   Engineering Societies Library
   345E 47th Street
   New York, New York 10017

2. U. S. Government Contract Reports:

   National Technical Information Service
   Springfield, Virginia 22151

3. IAEC reports

   International Atomic Energy Agency
   Kaerntnerring A 1010
   Vienna, Austria

   National Lending Library
   Boston Spa, England

4. Theses listed in Dissertation Abstracts + number (U.S.)

   University Microfilms
   Dissertation Copies
   P.O. Box 1764
   Ann Arbor, Michigan 48106

5. Patents: U. S. and Foreign

   Patent Office
   Box 9
   Washington, C. C. 20231

CONTENTS

I. BIBLIOGRAPHIES

1. Agajanian, A. H., "Semiconducting Devices: a bibliography of fabrication technology, properties, and applications", IFI/ Plenum, New York, 1976, p. 214-310, EEA80-36297.
2. Anon., "Ion solid interaction: a bibliography", Institute of Electrical Engineers, London, England, 1980.
3. Plunkett, J. C. and Stone, J. L., "A selected bibliography on ion implantation in solid state technology", Solid State Technol., 18 (12), 49-55 (1975), EEA79-18907.
4. Reimherr, G. W., "Ion implantation: a bibliography with abstracts; volume 1, 1964-August 1978", Report PB80-800105, NTIS, Springfield, VA, 1979, 301 pp.
5. Reimherr, G. W., "Ion implantation: a bibliography with abstracts; volume 2, September 1978-September 1979", Report PB80-800113, NTIS, Springfield, VA, 1979, 82 pp.
6. Reimherr, G. W., "Ion implantation: citations from NTIS data base, September 1978-September 1980", Report PB80-815996, NTIS, Springfield, VA, 1979, 82 pp.
7. Ting, R. N. and Subramanyam, K., "Ion implantation: an annotated bibliography", Report PB-245786, Sci. Eng. Libr., State Univ., New York, Buffalo, 1976, 218 pp., CA85-12551.
8. Zollars, G. F., "Ion implantation: citations from International Aerospace Abstracts data base, 1973-November 1979", Report PB80-803885, New Mexico University, Albuquerque, 1980, 55 pp.

II. BOOKS AND SYMPOSIA

9. Anderson, C. L., Celler, G. K. and Rozgonyi, G. A., (Eds.), "Laser and electron beam processing of electronic materials", Proceedings of the Symposium; The Electrochemical Society, Inc., Princeton, N. J., 1980.
10. Carter, G., Colligon, J. S. and Grant, W. A., (Eds.), Institute of physics conference series; No. 28: Applications of ion beams to materials, 1975, Publ. by Institute of Physics, London, England, 1976, CA84-158777.
11. Carter, G. and Grant, W. A., "Ion implantation of semiconductors", Halsted Press, New York, 1976, EEA79-32913.

12. Chernow, F., Borders, J. A. and Brice, D. K., (Eds.), 5th International Conference on Ion Implantation in Semiconductors, Held at: Boulder, Colorado, 8-13 August, 1976; Publ. by: Plenum Press, New York, 1977.
13. Gyulai, J., Lohner, T. and Pasztor, E., (Eds.), Proceedings of the International Conference on Ion Beam Modification of Materials, 1978, Budapest, Hungary.
14. Hirvonen, J. K., (Ed.), "Ion implantation", Academic Press, Inc., New York, 1980.
15. Mayer, J. W. and Rimini, E., (Eds.), "Ion beam handbook for material analysis", Academic Press, New York, 1977.
16. Meese, J. M., (Ed.), "Neutron transmutation doping in semiconductors", Plenum Press, New York, 1979, EEA83-9715.
17. Meyer, O., Linker, G. and Kappeler, F., "Ion beam surface layer analysis, Vol. 1 & 2", Plenum Press, New York, 1976.
18. Olson, C. L. and Schumacher, U., "Collective ion acceleration", Spring-Verlag, New York, 1979, EEA83-2271.
19. Perez, A. and Coussement, R., (Eds.), "Site characterization and aggregation of implanted atoms in materials", Plenum Press, New York, 1980.
20. Preece, C. M. and Hirvonen, J. K., (Eds.), "Ion implantation metallurgy", The Metallurgical Society of AIME, Warrendale, PA, 1980.
21. Saris, F. W. and Van der Weg, W. F., (Eds.), Proceedings of the 6th International Conference on Atomic Collision in Solids, Held at: Amsterdam, Netherlands, 1976, Publ. in Nucl. Inst. & Methods, Vol. 132, 1-732, (1976).
22. Scully, J. C., (Ed.), "Ion implantation and ion beam analysis techniques in corrosion studies", Pergamon Press, New York, 1977.
23. Smith, B., "Ion implantation range data for silicon and germanium device technologies", International School Book Service, P. O. Box 555, Forest Grove, Oregon, 1977.
24. Townsend, P. D., "Ion implantation, sputtering and their applications", Academic Press, New York, 1976, CA84-143719.
25. White, C. W. and Peercy, P. S., (Eds.),"Laser and electron beam processing of materials", Academic Press, New York, 1980.
26. Wilson, R. G. and Brewer, G. R., "Ion beams: with applications to ion implantation", R. E. Krieger Publishing Co., Inc., Huntington, New York, 1979.
27. Winterbon, K. B., "Ion implantation range and energy deposition distributions, Vol. 2: low incident ion energies", Plenum Press, New York, 1976, CA84-65559.
28. Wolicki, E. A., Butler, J. W. and Treado, P. A., Proceedings of the Third International Conference on Ion Beam Analysis, Held at: Washington, D. C., June 27-July 1, 1977; Publ. by North-Holland Publishing Company, New York, 1978.
29. Ziegler, J. F., (Ed.), "New uses of ion accelerators", Plenum Press, New York, 1975.

30. Proceedings of the AVS symposium on ion implantation - New prospects for materials modification, Publ. in: J. Vac. Sci. & Technol., Vol. 15, No. 5, 1978, PA82-14826.
31. International Conference on Low-Energy Ion Beams, Conference held at: Salford, England, September 5-8, 1977; Publ. by Institute of Physics, Bristol, England, 1978, EEA81-40203.
32. Proceedings of the Second International Conference on Ion Implantation Equipment, Conference held at: Provo-Trento, Italy, August 28-31, 1978; Publ. in Radiat. Eff., Vol. 44, No. 1-4, 1979, EEA83-14295.
33. Third International Conference on Ion Implantation: Equipment and Techniques, Abstracts, July 8-11, 1980, Queen's University, Kingston, Ontario, Canada.
34. Second International Conference on Ion Beam Modification of Materials, (Abstracts only), State University of New York, Albany, New York, July 14-18, 1980.

III. REVIEW ARTICLES

35. Alton, G. D., "Ion sources for accelerators", Conference on Applications of Small Accelerators, 3rd, 1974, p. 11-30, Publ. 1975, CA85-38296.
36. Andersen, H. H., "Sputtering and ion-source technology", IEEE Trans. Nucl. Sci., NS-23 (2), 959-66 (1976), CA84-172533.
37. Beanland, D. G., "The application of ion implantation to electronic device", Monitor, 38 (12), 191-9 (1977), EEA81-49725.
38. Bell, A. E., "Review and analysis of laser annealing", RCA Rev., 40 (3), 295-38 (1979), EEA83-9669.
39. Borders, J. A. and Arnold, G. W., "Applications of ion beam analysis to insulators", Report, Sandia Labs., Albuquerque, N. Mex., 1975, 19 pp., NSA32-27536.
40. Brown, W. L. and MacRae, A. U., "Ion implantation", Bell Lab. Rec., 53 (10), 388-94 (1975), PA79-10472.
41. Cachard, A. and Thomas, J. P., "Application of direct observation of nuclear reactions to the study of ion implantation", Acta Electron., 19 (1), 41-6 (1976), French, CA85-102812.
42. Chaumont, J., "Characteristics of an ion implantation set-up for research", Acta Electron., 19 (1), 23-8 (1976), French, PA79-66103.
43. Clark, D. J., "International conference on heavy ion sources: summary", IEEE Trans. Nucl. Sci., NS-23 (2), 1166-75 (1976), CA84-186181.
44. Datz, S., "Channeling: a tool for the study of the interactions of energetic particles penetrating solids", Radiat. Res., Proc. Int. Congr., 5th, 1974, p. 287-301, Publ. 1975, CA85-54674.

45. Dearnaley, G., "Ion bombardment and implantation", Surface Science. Lectures Presented at an International Course, Pt. II, 1974, p. 167-214, Publ. 1975, PA79-30959.
46. Dearnaley, G., "Ion implantation", Nature, 256 (5520), 701-5 (1975), PA78-76705.
47. Dearnaley, G., "Ion implantation for microelectronics", Proc. Int. Vac. Congr., 7th, 1977, 2, p. 1405-16, CA88-97808.
48. Donets, E. D., "Review of the JINR electron beam ion sources", IEEE Trans. Nucl. Sci., NS-23 (2), 897-903 (1976), EEA79-26273.
49. Dvurechensky, A. V., Kachurin, G. A., Mustafin, T. N. and Smirnov, L. S., "Laser annealing of ion-implanted semiconductors", AIP Conf. Proc., no. 50, 245-58 (1978), EEA82-34271.
50. Gibbons, J. F., "Applications of scanning CW lasers and electron beams in silicon technology", Jpn. J. Appl. Phys., 19, suppl. 19-1, 121-8 (1979), EEA83-48730.
51. Goetzberger, A. and Frizsche, C. R., "Ion implantation in semiconductors", Ber. Kernforschungsanlage Juelich (Conf.), 69-73 (1976), German, CA86-36748.
52. Grant, W. A. and Navinsek, B., "Some current applications of ion implantation", 8th International Summer School on the Physics of Ionized Gases, 1976, p. 329-39, PA80-43236.
53. Grant, W. A. and Williams, J. S., "The modification of surface layers by ion implantation", Sci. Prog., 63 (249), 27-64 (1976), CA84-96754.
54. Guernet, G., "Development of the characteristics and performance of implantation machines", Acta Electron., 19 (1), 13-22 (1976), French, CA85-102810.
55. Guernet, G., "Developments in ion implantation techniques provide new and improved electronic components and materials", Electron. and Appl. Ind., no. 238, 15-18 (1977), French, EEA81-10266.
56. Gyulai, J., "Ion implantation research", Fiz. Sz., 26 (11), 423-7 (1976), Hungarian, EEA80-24864.
57. Gyulai, J., "Ion implantation induced defects and epitaxial regrowth by thermal and laser annealing", International Conference on Defects and Radiation Effects in Semiconductors, 1978, p. 128-47, PA83-24479.
58. Hara, T., "Trends in ion implantation in gallium arsenide", Solid State Technol., 22 (11), 69-74 (1979), EEA83-19005.
59. Hartley, N. E. W., "Friction and wear of ion-implanted metals - a review", Thin Solid Films, 64 (2), 177-90 (1979), PA83-40555.
60. Hemment, P. L. F., "Ion implantation into compound semiconductors", Applications of Ion Beams to Materials, 1975, p. 44-58, Publ. 1976, PA79-69556.
61. Hemment, P. L. F., "Dosimetry techniques (applicable to ion implantation)", Radiat. Eff., 44 (1-4), 31-46 (1979), EEA83-14297.

62. Hlavka, J., "Ion implantation: possibilities and exploitation", Cesk. Cas. Fis. A, 25 (2), 128-38 (1975), Czech, EEA78-37386.
63. Hofker, W. K., "Physical aspects of ion implantation", Acta Electron., 19 (1), 29-40 (1976), CA85-71106.
64. Krynicki, J., "Annealing of ion implanted layers by laser beam", Electron Technol., 12 (1), 77-90 (1979), EEA83-9693.
65. Liau, Z. L. and Mayer, J. W., "Limits of composition achievable by ion implantation", J. Vac. Sci. and Technol., 15 (5), 1629-35 (1978), PA82-14827.
66. MacRae, A. U., "Ion implantation in silicon: research and applications", J. Vac. Sci. & Technol., 14 (1), 531 (1977), EEA80-28827.
67. Mark, M., "Doping of semiconductors by ion implantation", AMR, no. 7-8, 221-3 (1975), German, EEA79-4534.
68. Masnari, N., "Ion implantation in semiconductor technology", Semicond. Electron. Integr. Circuits, Course Notes, Eng. Summer Conf., 1978, 51 pp., CA91-100368.
69. Mazzoldi, P., "Ion implantation: a fact and applications", Tecnol. Elettr., 11, 56-65 (1977), Italian, EEA81-33280.
70. McKenna, C. M., "High current dosimetry techniques (for ion implantation)", Radiat. Eff., 44 (1-4), 93-110 (1979), EEA83-14301.
71. Meese, J. M., "The NTD process - a new reactor technology", Neutron Transmutation Doping in Semiconductors, 1978, Publ. Plenum 1979, p. 1-10, EEA83-9716.
72. Meese, J. M., Cowan, D. L. and Chandrasekhar, M., "A review of transmutation doping in silicon", IEEE Trans. Nucl. Sci., NS-26 (6), 4858-67 (1975), EEA81-21325.
73. Middleton, R., "A review of sputter negative ion sources", IEEE Trans. Nucl. Sci., NS-23 (2), 1098-103 (1976), CA85-11256.
74. Middleton, R., "A review of negative ion sources", Rev. Phys. Appl., 12 (10), 1435-44 (1977), EEA81-21325.
75. Namba, S., "Microfabrication techniques for VLSI and other devices", Proceedings of the International Conference on Ion Beam Modification of Materials, 1978, p. 887-914.
76. Nicholas, K. H., "Semiconductor processing", Radiat. Eff., 44 (1-4), 111-20 (1979), EEA83-14302.
77. Pattyn, H., "Ion implantation in ferromagnets", Verh. K. Acad. Wet. Lett. and Schone Kunsten Belg., 39 (141), 5-72 (1977), Flemish, PA81-79623.
78. Pfister, J. C., "Ion implantation", Radiation Damage Processes in Materials, 1973, p. 467-75, Publ. 1975, EEA79-4595.
79. Pickar, K. A., "Ion implantation in silicon: Physics, processing, and microelectronic devices", In: Applied solid state science: Advances in materials and device research, Vol. 5, Wolfe, R., (Ed.), Academic Press, New York, p. 151-249 (1975), CA84-98174.
80. Picraux, S. T., "Implantation metallurgy", Report SAND-75-5515, Sandia Lab., Albuquerque, N. Mex., 1975, 24 pp., CA84-183467.

81. Rimini, E., Baeri, P., Campisano, S. U. and Foti, G., "Ruby laser pulse effects in ion implanted semiconductors", AIP Conf. Proc., no. 50, 259-73 (1978), EEA82-34272.
82. Rupprecht, H. S., "New advances in semiconductor implantation", J. Vac. Sci. & Technol., 15 (5), 1669-74 (1978), EEA82-8266.
83. Rupprecht, H. S., Van de Wiele, F., Engl, W. L. and Jespers, P. G., "Ion implantation", Proceedings of the NATO Advanced Study Institute on Process and Device Modelling for Integrated Circuit Design, 1977, p. 115-39, EEA81-38130.
84. Ryssel, H., "The impact of ion implantation on VLSI", 1978 International Electron Devices Meeting, p. 583-7, EEA83-4630.
85. Sansbury, James, "Applications of ion implantation in semiconductor processing", Solid State Technol., 19 (11), 31-7, 43 (1976), CA86-36745.
86. Schmidt, A., "Use of magnetic laboratory isotope separators for ion implantation in ZfK Rossendorf", Isotopenpraxis, 12 (3), 107-14 (1976), German, CA84-171010.
87. Shepherd, D. M., "A review of some suppressed accelerator tube installations", Rev. Phys. Appl., 12 (10), 1521-4 (1977), EEA81-21288.
88. Smith, T. C., "Silicon production applications of ion implantation", IEEE Trans. Nucl. Sci., NS-26 (1), 1677-82 (1979), EEA82-26749.
89. Stephens, K. G. and Wilson, I. H., "Properties and applications of ion-implanted films", Thin Solid Films, 50, 325-47 (1978), EEA81-42359.
90. Stevens, B. A., "A bibliography on laser processing of semiconductors", AIP Conf. Proc., no. 50, 671-82 (1978), EEA82-38180.
91. Stritzker, B., "Ion implantation and superconductivity", J. Nucl. Mater., 72 (1-2), 256-62 (1977), EEA81-28790.
92. Titov, V. V., "Ion implantation - problems and perspectives", Phys. Status Solidi A, 48 (1), 13-22 (1978), EEA81-42356.
93. Tokuyama, T. "Use of ion implantation in device fabrication at Hitachi CRL", In: Ion Implantation Semicond. (Proc. Int. Conf. Ion Implantation Semicond. Other Mater.), Chernow, F., Borders, J. A. and Brice, D. K. (Eds.), Plenum, New York, N. Y., 519-33 (1977), CA89-83337.
94. Townsend, P. D., "Ion implantation in optical materials", Applications of Ion Beams to Materials, 1975, p. 104-11, Publ. 1976, PA79-68592.
95. Townsend, P. D., "Some applications of ion implantation to colour centre research", Radiat. Phys. and Chem., 11 (5), 205-10 (1978), PA82-19272.
96. Van der Weg, W. F. and Tamminga, Y., "High energy ion scattering as a tool for studying ion implantation", Acta Electron., 19 (1), 47-52 (1976), CA86-71107.
97. Wilson, R. G., "Intense ion beams for implantation", International Conference on Low-Energy Ion Beams, 1977, p. 50-62, EEA81-40210.

98. Wittkower, A. B., Rose, P. H. and Ryding, G., "Advances in ion implantation production equipment", Solid State Technol., 18 (12), 41-5 (1975), EEA79-8786.

IV. THEORY

99. Akkerman, A. F., "Computer simulation of the range and defects spatial distributions for low kiloelectronvolt light ion implantation of solids", Proceedings of the International Conference on Ion Beam Modification of Materials, 1978, p. 45-54.
100. Anttila, A. and Hautala, M., "Radiation enhanced outdiffusion during ion implantation", Appl. Phys., 19 (2), 199-203 (1979), PA82-73200.
101. Arandjelovic, V., Miljkovic, L. and Tjapkin, D., "The numerical analysis of anomalous doping profiles of phosphorus in silicon", Solid-State Electron., 22 (4), 355-9 (1979), EEA82-38257.
102. Chu, W. K., Moruzzi, V. L. and Ziegler, J. F., "Calculations of the energy loss of ^{4}He ions in solid elements", J. Appl. Phys., 46 (7), 2817-20 (1975), PA78-72502.
103. Collins, R. and Carter, G., "Spatial distribution of ions implanted into solids subject to diffusion and surface sputtering", Radiat. Eff., 26 (3), 181-91 (1975), CA84-52891.
104. Combasson, J. L., "Mathematical models of ion implantation", Vide, 171, 321-8 (1974), French, CA86-64084.
105. Cooper, R. K., "Solenoid-lens effects in beam-transport equations", Part. Accel., 7 (1), 41-7 (1975), EEA79-14466.
106. Desalvo, A. and Rosa, R., "Monte Carlo calculations on spatial distribution of implanted ions in silicon", Radiat. Eff., 31 (1), 41-5 (1976).
107. Desalvo, A. and Rosa, R., "A comprehensive computer program for ion penetration in solids", Proceedings of the International Conference on Ion Beam Modification of Materials, 1978, p. 127-36.
108. Finger, U., Gartner, K. and Hehl, K., "Calculation of axial dechannelling", Wiss. Z. Friedrich-Schiller Univ. Jean, Math. Naturwiss. Reihe, 24 (5-6), 541-6 (1975), German, PA79-61391.
109. Freeman, J. H. and Booker, D. V., "Ion beam studies. V. The computer simulation of composite ion implantation profiles", Nucl. Instrum. & Methods, 144 (2), 175-81 (1977), EEA80-41392.
110. Gries, W. H., "Quantitative ion implantation: theoretical aspects", Int. J. Mass Spectrom. and Ion Phys., 30 (2), 97-112 (1979), PA82-69140.
111. Hall, B. O., "Stress distribution in helium-ion implantations", J. Nucl. Mater., 63, 285-91 (1976), PA80-35599.
112. Jackson, D. P. and Morgan, D. V., "Computer modelling of planar dechannelling, I: Monatomic lattices", Radiat. Eff., 28 (1-2), 5-13 (1976), PA84-187720.

113. Kalbitzer, S. and Oetzmann, H., "Ranges and range theories", Proceedings of the International Conference on Ion Beam Modification of Materials, 1978, p. 3-34.
114. Kraeutle, H., "Model calculations of profiles and dose of high dose ion implants influenced by sputtering", Nucl. Instrum. Methods, 134 (1), 167-72 (1976), CA85-55280.
115. Krimmel, E. F., "Maximum range of axial channelled ions", Phys. Lett. A, 53A (3), 231-2 (1975), PA78-58089.
116. Kumakhov, M. A., "The theory of channeling at small depths", Radiat. Eff., 26 (1-2), 43-8 (1975), PA79-13545.
117. Matsumura, H. and Furukawa, S., "Theoretical considerations in lateral damage distribution formed by ion-implantation", Jap. J. Appl. Phys., 14 (11), 1783-90 (1975), PA79-13536.
118. Matsumura, H. and Furukawa, S., "Lateral spread of damage formed by ion implantation", J. Appl. Phys., 47 (5), 1746-51 (1976), PA79-66115.
119. Morgan, D. V. and Jackson, D. P., "Computer modelling of planar dechanneling. II. Diatomic lattices", Radiat. Eff., 29 (2), 99-106 (1976), PA79-68583.
120. Morgan, D. V. and Jackson, D. P., "Computer modelling of planar dechanneling. III. Trajectory-dependent electron multiple scattering", Radiat. Eff., 29 (2), 107-11 (1976), PA79-68584.
121. Morozov, N. P., Tetelbaum, D. I., Pavlov, P. V. and Zorin, E. I., "The calculation of secondary defect formation of ion implantation of silicon", Phys. Status Solidi A, 37 (1), 57-64 (1976), PA79-90857.
122. Nicholas, K. H., "A rapid method of predicting the sheet resistances of boron implanted layers", Radiat. Eff., 28 (3-4), 177-81 (1976), PA79-62374.
123. Pagel, L. and Scheit, U., "Calculating field potential of ion paths and plasma boundaries in ion source's extraction system", Wiss. Z. Tech. Hochsch. Karl-Marx-Stadt, 17 (3), 349-54 (1975), German, EEA79-14474.
124. Smith, B. J. and Jain, A., "Electronic stopping cross sections for use in ion range calculation (semiconductor doping)", Report AERE-R8326, UKAEA, Harwell, Oxon., England, 1976, 7 pp., EEA79-41427.
125. Tsong, I. S. T., Corredor, A., White, W. B., Tolk, N. H. and Kraus, J. F., "An estimation of the quantum yield in the cando-luminescence process by low energy H^+ ion bombardment", J. Electrochem. Soc., 125 (12), 2015-19 (1978).
126. Winterbon, K. B., "Calculations of implanted-ion range and energy-deposition distributions: ^{11}B in Si", Radiat. Eff., 30 (4), 199-206 (1976).
127. Winterbon, K. B., "Ion implantation distributions in inhomogeneous materials", Appl. Phys. Lett., 31 (10), 649-51 (1977), EEA81-10260.

128. Winterbon, K. B., "Ion-implantation distributions in non-uniform targets: projected range", Radiat. Eff., 39 (1), 31-8 (1978), PA82-23625.
129. Winterbon, K. B., "On the theory of recoil implantation", Proceedings of the International Conference on Ion Beam Modification of Materials, 1978, p. 35-44.
130. Winterbon, K. B. and Sanders, J. B., "Analytical calculations of some ion-implantation depth distributions", Radiat. Eff., 39 (1), 39-44 (1978), PA82-23626.
131. Winterbon, K. B. and Sanders, J. B., "Analytical calculations of some ion implantation depth distributions. II. Inhomogeneous equations", Radiat. Eff., 46 (1-2), 15-22 (1980), PA83-84203.
132. Wood, R. F., Giles, G. E. and Kirkpatrick, J. R., "Theoretical analysis of thermal and mass transport during laser annealing", Proc. Soc. Photo-Opt. Instrum. Eng., 198, 2-10 (1979), EEA83-48707.
133. Zaremba, A. J., Marcyk, G. T. and Streetman, G. G., "Optimal summation of gaussians for ion implantation profile control", IEEE Trans. Electron Devices, ED-24 (2), 163-5 (1977), EEA80-6576.

V. MATERIALS

1. General

134. Abroyan, I. A., Titov, A. I. and Khlebalkin, A. V., "Formation of a surface peak of structure defects formed by ion irradiation", Sov. Phys.-Semicond., 11 (6), 712-13 (1977), EEA81-10137.
135. Anderson, C. L., Dunlap, H. L. and Vaidyanathan, K. V., "Ion implantation of wide bandgap semiconductors", Report AD-A062367, Hughes Research Labs, Malibu, CA, 1978, 21 pp.
136. Anon., "Ion implantation and its applications to semiconductors", Ing. & Tech., no. 302, 33-8 (1976), French, EEA79-28397.
137. Anon., "Ion implantation as a new surface treatment technology", Report PB-296966/5SL, National Materials Advisory Board (NRC), Washington, DC, 1979, 53 pp.
138. Appleton, B. R., Larson, B. C., White, C. W., Narayan, J., Wilson, S. R. and Pronko, P. P., "New materials properties achievable by ion implantation doping and laser processing", AIP Conf. Proc., no. 50, 291-8 (1978), EEA82-34274.
139. Asche, W., "Nuclear transmutation as a method of doping in semiconductor technology", Funkschau, 50 (9), 397-8 (1978), German, EEA81-49717.
140. Auston, D. H., Golovchenko, J. A., Simons, A. L., Slusher, R. E., Smith, P. R. and Surko, C. M., "Dynamics of laser annealing", AIP Conf. Proc., no. 50, 11-26 (1978), EEA82-34265.

141. Auston, D. H., Golovchenko, J. A., Simons, A. L. and Surko, C. M., "Dynamics of Q-switched laser annealing", Appl. Phys. Lett., 34 (11), 777-9 (1979), EEA82-38215.
142. Auston, D. H., Golovchenko, J. A. and Venkatesan, T. N. C., "Dual-wavelength laser annealing", Appl. Phys. Lett., 34 (9), 558-60 (1979), EEA82-34263.
143. Auston, D. H., Golovchenko, J. A. and Venkatesan, T. N. C., "Dynamics of laser annealing", Proc. Soc. Photo-Opt. Instrum. Eng., 198, 19-25 (1979), EEA83-48709.
144. Bayer, E. H., Kranik, J. R. and Mueller, W. F., "Heat transfer 'pillow' for use in high current ion implantation", IBM Tech. Disclosure Bull., 22 (9), 4073 (1980), EEA83-39747.
145. Bayerl, P., Ryssel, H. and Ramin, M., "High energy implantation of buried insulating layers", Radiat. Eff., 47 (1-4), 217-20 (1980), EEA83-43990.
146. Behrendt, R., "Adjustment of operating voltage of avalanche devices by suitable implantation doping", Int. Conf. Ion Implantation Semicond., Proc., 1978, p. 590-5, CA90-96422.
147. Bernstein, T., "A useful method for approximating the profile of ions implanted through a thin film", IEEE Trans. Electron Devices, ED-24 (12), 1365-6 (1977), EEA81-10269.
148. Bhattacharya, R. S., Kerkdijk, C. B., Van Der Veen, J. F. and Saris, F. W., "Photon emission studies of the synthesis of molecules by ion implantation", Radiat. Eff., 33 (1), 57-8 (1977), PA80-70297.
149. Bilyus, K. I. and Pranyavichyus, L. I., "Influence of sputtering on the distribution profile of ion-implanted impurities", Sov. Phys.-Semicond., 12 (7), 801-2 (1978), EEA82-22092.
150. Blamires, N. G. and Smith, B. J., "The range of doubly charged boron ions", J. Phys. D, 10 (5), 799-804 (1977), EEA80-20917.
151. Bond, N. R., Doney, D. A. and Makris, J. S., "Improved leakage-limited yield using an anneal", IBM Tech. Disclosure Bull., 22 (3), 548-9 (1979), EEA83-9694.
152. Borders, J. A., "Introduction to several solid state techniques for the study of ion implanted materials", Report SAND-78-0870C, Sandia Labs, Albuquerque, NM, Department of Energy, 1978, 25 pp.
153. Brack, K. and Schmitt, A., "Directed activation of implanted semiconductor zones", IBM Tech. Disclosure Bull., 21 (3), 1086 (1978), EEA82-17404.
154. Bradley, Kenneth Reed, "Laser annealing of and laser interactions with ion implant semiconducting materials", Report AD-A080363/5, Air Force Inst. of Tech. Wright-Patterson, AFB, OH, School of Engineering, 1979, 74 pp.
155. Brice, D. K., "Ion implantation range and energy deposition codes corel, rase4, and damg2", Report SAND-75-0622, Sandia Labs, Albuquerque, N. Mex., 1977, 115 pp.
156. Briska, M. and Schackert, K., "Treating thin films by pulsed ion implantation", IBM Tech. Disclosure Bull., 22 (9), 4107 (1980).

157. Brown, W. L., Golovchenko, J. A., Jackson, K. A., Kimerling, L. C., Leamy, H. J., Miller, G. L., Poate, J. M., Rodgers, J. W., Rozgonyi, G. A., et al., "Laser annealing of ion-implanted semiconductors", Rapid Solidification Process: Princ. Technol. Proc. Int. Conf., 1977, p. 123-8, CA91-31534.
158. Bruel, M., Berthet, B., Floccari, M. and Michaud, J. F., "Target heating during ion implantation and related problems", Radiat. Eff., 44 (1-4), 173-80 (1979), EEA83-14304.
159. Carruthers, R., Nagasaki, T., Poponiak, M. R. and Zielinski, L. B., "Removable high temperature ion implant mask", IBM Tech. Disclosure Bull., 18 (4), 1214 (1978).
160. Carter, G. and Collins, R., "The spatial distribution of ions implanted into solids subject to diffusion, local trapping and surface sputtering", Radiat. Eff., 28 (1-2), 123-5 (1976), PA79-53239.
161. Christodoulides, C. E., Grant, W. A. and Williams, J. S., "The determination of high-dose implantation profiles using low-angle Rutherford backscattering", Nucl. Instrum. & Methods, 149 (1-3), 219-24 (1978), EEA82-805.
162. Chu, W., "Very low dose ion implantation by scattering method", IBM Tech. Disclosure Bull., 21 (8), 3188-9 (1979), EEA82-38247.
163. Chu, W. K., Poponiak, M. R., Alessandrini, E. I. and Lever, R. F., "Interference of arsenic diffusion by argon implantation", Proceedings of the International Conference on Ion Beam Modification of Materials, 1978, p. 313-28; Also: Radiat. Eff., 49, 23-28 (1980).
164. Colligon, J. S. and Fuller, D., "Secondary ion emission studies of the range profiles of implanted ions", Radiat. Eff., 28 (3-4), 183-7 (1976), CA85-27783.
165. Collins, R. J. and Peria, W. T., "Computer simulation of pulsed laser and electron beam annealing and doping of semiconductors", Proc. Soc. Photo-Opt. Instrum. Eng., 198, 11-18 (1979), EEA83-48708.
166. Davies, J. A., "Interaction of ion beams with solids", J. Vac. Sci. & Technol., 12 (1), 438 (1975), PA79-1679.
167. De La Moneda, F. H., "Ion implanted layers with abrupt edges", Patent US 4,202,002, Publ. May 1980.
168. Dearnaley, G., "Atomic-displacement mechanism in ion-bombarded semiconductors and insulators", Appl. Phys. Lett., 26 (9), 499-501 (1975), PA78-58085.
169. Dinklage, J. B., Keenan, W. A. and Tice, W. K., "Gettering by ion implantation", IBM Tech. Disclosure Bull., 19 (11), 4091 (1977), EEA81-10270.
170. Dvurechensky, A. V., Kachurin, G. A. and Antonenko, A. Kh., "The mechanisms of impurity redistribution on laser-annealing of ion-implanted semiconductors", Radiat. Eff., 37 (3-4), 179-81 (1978), EEA81-45940.

171. Dvurechensky, A. V., Kachurin, G. A., Mustafin, T. N. and Smirnov, L. S., "Laser annealing of ion-implanted semiconductors", AIP Conf. Proc., no. 50, 245-58 (1978), EEA82-34271.
172. Dvurechensky, A. V. and Smirnov, L. S., "Production and rearrangement of radiation defects in ion implanted semiconductors", Radiat. Eff., 37 (3-4), 173-8 (1978), CA89-189808.
173. Eernisse, E. P., "Use of ion implantation for radiation damage studies", Proceedings of the 4th Conference on the Scientific and Industrial Applications of Small Accelerators, 1977, p. 598-601, PA80-80530.
174. Elliott, D. J., "Positive photoresists as ion implantation masks", Proceedings of the Society of Photo-Optical Instrumentation Engineers, 1979, p. 153-72, EEA83-14316.
175. Emmanuel, J. C., Rochet, D. and Schwerer, A., "Aluminum silicon coating for reducing metallic sputtered impurities in implanted semiconductor wafers", IBM Tech. Disclosure Bull., 22 (9), 4094 (1980), EEA83-39749.
176. Ennis, R. M., Jr., "Combined ion implantation and kinetic transport deposition process", Patent USA 3908183, Publ. September 1975, PA79-18910.
177. Fortino, A. G. and Geipel, H. J., "Ion implant uniformity measuring technique", IBM Tech. Disclosure Bull., 22 (4), 1401-2 (1979), EEA83-14285.
178. Foti, G., "Laser pulse annealing in semiconductors", Proceedings of the International Conference on Ion Beam Modification of Materials, 1978, p. 611-28.
179. Foti, G., Baeri, P., Rimini, E. and Campisano, S. U., "Off-axis channeling disorder analysis", J. Appl. Phys., 47 (12), 5206-13 (1976), EEA80-13453.
180. Freeman, J. H., "The boundaries of ion implantation", Applications of Ion Beams to Materials, 1975, p. 340-56, Publ. 1976, EEA79-32928.
181. Freeman, J. H. and Booker, D. V., "Ion beam studies. V. The computer simulation of composite ion implantation profiles", Nucl. Instrum. and Methods, 144 (2), 175-81 (1977), EEA80-41392.
182. Freeman, J. H. and Temple, W., "The ion implantation of powders", Radiat. Eff., 28 (1-2), 85-6 (1976), PA79-53238.
183. Friedland, E., "Influence of straggling on the analysis of ion implantation profiles using backscattering or nuclear reactions", Nucl. Instrum. Methods, 150 (2), 301-4 (1978), EEA81-47401.
184. Fuller, C. T., "Variable-temperature sample system for ion implantation at -192 to +500°C", Report SAND-78-0520, Sandia Lab., Albuquerque, N. Mex., 1978, 17 pp., CA90-15147.
185. Gailliard, J. P., Dupuy, M., Garcia, M., Roussin, J. C. and Roche, M., "High dose implantations of antimony for buried layer applications", Radiat. Eff., 47 (1-4), 131-5 (1980), EEA83-44014.

186. Gailliard, J. P. and Robie, J. Y., "Some applications of ion implantation in materials", Acta Electron., 19 (2), 169-78 (1976), French, PA79-65145.
187. Gardner, J. A., Lever, R. F. and Michel, A. E., "Ion implant heat transfer", IBM Tech. Disclosure Bull., 21 (11), 4526 (1979), EEA83-737.
188. Geipel, H. J. and Tice, W. K., "Critical microstructure for ion-implanted gettering", IBM Tech. Disclosure Bull., 19 (11), 4092-3 (1977), EEA81-10271.
189. Gibbons, J. F., "Ion implantation and proton-enhanced diffusion in semiconductors", J. Vac. Sci. & Technol., 12 (1), 475 (1975), EEA79-836.
190. Grant, W. A. and Williams, J. S., "The modification of surface layers by ion implantation", Sci. Prog., 63 (249), 27-64 (1976), PA79-61387.
191. Greenstein, E. and MacIver, B. A., "Selective enhancement of phosphorus diffusion by implanting halogen ions", Patent USA 4,133,701, Publ. January 1979, CA90-113974.
192. Greenwald, A. C., Kirkpatrick, A. R., Little, R. G. and Minnucci, J. A., "Pulsed-electron-beam annealing of ion-implantation damage", J. Appl. Phys., 50 (2), 783-7 (1979), EEA82-30240.
193. Gries, W. H., "Quantitative ion implantation: the practice", Int. J. Mass Spectrom. and Ion Phys., 30 (2), 113-25 (1979), PA82-69141.
194. Guernet, G., "Developments in ion implantation techniques provide new and improved electronic components and materials", Electron. & Appl. Ind., no. 238, 15-18 (1977), French, EEA81-10266.
195. Haas, E., Hofmann, R. and Glawischnig, H., "Removal of impurities originating in semiconductor materials during ion implantation processes", Patent Germany 2709802, Publ. September 1978, CA89-172611.
196. Haas, E. W., "Error due to isotope composition in analytical investigation of dopant implantation", Radiochem. & Radioanal. Lett., 41 (2), 119-28 (1979), German, EEA83-14307.
197. Haas, E. W. and Martin, J. A., "Nuclear transmutation doping from the viewpoint of radioactivity formation", Neutron Transmutation Doping in Semiconductors, 1978, p. 27-36, EEA83-9718.
198. Hake, D. S., "Ion implantation impurity analysis by glow discharge optical spectroscopy", Report AD-A034007, Sch. Eng., Air Force Inst. Technol., Wright-Patterson AFB, Ohio, 1976, 84 pp., CA86-198789.
199. Hammer, W. N., Michel, A. E., "A technique for measuring, displaying, recording, and modifying the spatial uniformity of implanted ions", J. Appl. Phys., 47 (5), 2161-4 (1976), EEA79-32849.
200. Harada, H., "Boron and phosphorus ion-implantation profiles by junction depth measurements", Oyo Buturi, 45 (10), 995-8 (1976), Japanese, EEA80-17174.

201. Hemment, P. L. F., "Absolute dosimetry of ion implanted impurities using a calorimetric method", Vacuum, 27 (10-11), 611-16 (1977), EEA81-14662.
202. Hendrickson. T. E., "Ion implanted profiles from two point spreading resistance measurements", J. Electrochem. Soc., 122 (11), 1539-41 (1975), PA79-14700.
203. Hess, L. D. and Yaron, G., "Laser annealing [semiconductor processing]", Ind. Res./Dev., 21 (11), 141, 143-4, 146-7, 149, 152 (1979), EEA83-39772.
204. Heywang, W., Krimmel, E. F. and Runge, H., "Annealing mechanism of radiation damage and dopants in pulsed laser light irradiated ion implanted layers", Phys. Status Solidi A, 51 (1), K79-82 (1979), EEA82-26751.
205. Hirvonen, J. K., Butler, J. W., Smith, T. P., Kant, R. A. and Westcott, V. C., "Sliding-wear reduction by ion implantation", Radiat. Eff., 49 (1-3), 73-4 (1980), PA84-3938.
206. Hodgson, R. T., "Pulse ion implant and annealing", IBM Tech. Disclosure Bull., 23 (1), 373 (1980), PA84-1282.
207. Hlavka, J. and Rybak, M., "Sensitive detection of surface inhomogeneities", Scr. Fac. Sci. Nat. Univ. Purkynianae Brunensis Phys., 4 (2), 53-8 (1974), EEA79-8797.
208. Hofker, W. K., "Physical aspects of ion implantation", Acta Electron., 19 (1), 29-40 (1976), PA79-66104.
209. Hu, W.-W., Clayton, C. R. and Herman, H., "Fatigue-life enhancement by ion implantation", Scr. Metall., 12 (8), 697-98 (1978).
210. Jamba, D. M., "Some aspects of dose measurement for accurate ion implantation", Report NBS-SP-400-39, Nat. Bur. Stand., Washington, D. C., USA, 1977, 36 pp., EEA81-1207.
211. Kachurin, G. A. and Nidaev, E., "Laser annealing of radiation defects in low disordered layers", Proceedings of the International Conference on Ion Beam Modification of Materials, 1978, p. 719-26.
212. Kachurin, G. A., Nidaev, E. V., Khodyachikh, A. V. and Kovaleva, L. A., "Annealing of implanted layers by a scanning laser beam", Fiz. Tekh. Poluprovodn, 10 (10), 1890-3 (1976), Russian, CA85-201102.
213. Kachurin, G. A., Pridachin, N. B. and Smirnov, L. S., "Annealing of radiation defects by laser radiation pulses", Sov. Phys.-Semicond., 9 (7), 946 (1975), EEA79-20327.
214. Kamoshida, M., "Ion implantation from the view point of micro fabrication for silicon devices", J. Jpn. Soc. Precis. Eng., 44 (2), 236-42 (1978), Japanese, EEA81-45937.
215. Kanamori, A., "Reduction in minority carrier storage effect by flourine ion implantation damage", IEEE trans. Electron Devices, ED-25 (7), 772-8 (1978), EEA81-38169.
216. Kaplan, R. and Wagner, R. J., "Cyclotron resonance characterization of ion-implanted carriers in semiconductors", J. Vac. Sci. & Technol., 13 (4), 899-902 (1976), EEA80-9294.

217. Kedro, M. and Bezak, V., "Longitudinal electrical conductivity of semiconducting films with an inverse layer due to implanted ions", Thin Solid Films, 36 (1), 113-16 (1976), EEA79-41315.
218. Kelly, R. and Sanders, J. B., "On the role of recoil implantation in altering the stoichiometry of a bombarded solid", Nucl. Instrum. Methods, 132, 335-43 (1976), CA85-27723.
219. Kiriakidis, G., Christodoulides, C. E., Carter, G. and Colligon, J. S., "An RBS technique for measurement of the erosion rate of ion implanted films", Appl. Phys., 19 (2), 191-4 (1979), PA82-74038.
220. Kirkpatrick, A. R., Minnucci, J. A., Greenwald, A. C. and Little, R. G., "Transient processing of implanted semiconductors by pulsed electron beams", Proceedings of the International Conference on Ion Beam Modification of Materials, 1978, p. 629-52.
221. Komatsubara, K., "Ion-implantation into compound semiconductors", Oyo Buturi, 45 (3), 194-205 (1976), PA79-69543.
222. Kondo, M., "Method of implanting ions of different mass numbers in semiconductor crystals", Patent USA 3897276, Publ. July 1975.
223. Koutnik, M. and Rysava, M., "Investigation of ion implanted materials by in-depth Auger analysis", Proceedings of the International Conference on Ion Beam Modification of Materials, 1978, p. 599-610.
224. Krynicki, J., "Annealing of ion implanted layers by laser beam", Electron Technol., 12 (1), 77-90 (1979), EEA83-9693.
225. Kudo, H., "Dechannelling of fast ions in distroted crystals. I. Dislocations", J. Phys. Soc. Jap., 40 (6), 1645-53 (1976), PA79-65139.
226. Land, D. J., Simons. D. G., Brennan, J. G., Brown, M. D., Hirvonen, J. K. and Biersack, J. P., "Range distributions for 25-200 keV $^{14}N^{+}$ ions", Proceedings of the International Conference on Ion Beam Modification of Materials, 1978, p. 93-102.
227. Lecrosnier, D. P., Pelous, G. P. and Burgeat, J., "Investigation of ion-implantation damage and X-ray double reflection", Appl. Phys. Lett., 30 (3), 141-3 (1977), EEA80-9393.
228. Lecrosnier, D., Pelous, G. and Richou, F., "Long range enhancement of boron diffusivity by phosphorus diffusion", Proceedings of the International Conference on Ion Beam Modification of Materials, 1978, p. 421-28.
229. Liau, Z. L. and Mayer, J. W., "Limits of composition achievable by ion implantation", J. Vac. Sci. Technol., 15 (5), 1629-35 (1978).
230. Look, D. C., Meese, J. M., Kreitman, M. M. and Manthuruthil, J. C., "Identification of defects in compound semiconductors", Report AD-A032128/1SL, Dayton Univ., Ohio, Dept. of Physics, 1976, 217 pp.

231. Lurio, A. and Ziegler, J. F., "X-rays produced during ion implantation", Appl. Phys. Lett., 31 (7), 482-4 (1977), EEA81-5913.
232. Magee, C. W. and Wu, C. P., "Hydrogen ion implantation profiles as determined by SIMS", Nucl. Instrum. & Methods, 149 (1-3), 529-33 (1978), EEA82-808.
233. Malin, K. and Schmitt, A., "Measuring temperature occurring during ion implantation", IBM Tech. Disclosure Bull., 23 (3), 1096 (1980).
234. Marcyk, G. T., "Atomic and electrical profile studies of ion-implanted semiconductors", Report AD-A069782, Illinois Univ. at Urbana-Champaign, 1978, 127 pp.
235. Martin, P. J. and MacDonald, R. J., "Characteristic photon emission from ion bombarded silicon and silica surfaces", Phys. Lett. A, 55A (8), 483-5 (1976), CA84-128162.
236. Masters, B. J., "Proton implantation process for reducing alpha-induced 'soft errors'", IBM Tech. Disclosure Bull., 22 (12), 5326 (1980), EEA83-48697.
237. Masuda, K., "Heavy-dose ion implantation", International Conference on Radiation Effects in Semiconductors, 1977, p. 174-85, EEA80-41404.
238. Matsumoto, H., "Impurity doping of silicon by thermal neutron irradiation", Annu. Rep. Eng. Res. Inst. Fac. Eng. Univ. Tokyo, 37, 123-6 (1978), Japanese, EEA82-22079.
239. Matsumura, H., "Lateral spread of implanted ions", Oyo Buturi, 47 (6), 508-21 (1978), Japanese, EEA82-4306.
240. Matsumura, H. and Furukawa, S., "Lateral spread of damage formed by ion implantation", J. Appl. Phys., 47 (5), 1746-51 (1976), CA85-39838.
241. Matteson, S., Tonn, D. G. and Nicolet, M. A., "Diagnostic test for ion implantation dosimetry", J. Vac. Sci. & Technol., 16 (3), 882-3 (1979), EEA82-47081.
242. McMahon, R. A. and Ahmed, H., "Electron-beam annealing of ion implanted silicon", Electron. Lett., 15 (2), 45-7 (1979), CA90-131707.
243. Miyao, M. and Tokuyama, T., "Laser annealing of ion implanted semiconductors", Oyo Buturi, 48 (7), 664-70 (1979), Japanese, EEA83-9702.
244. Monfret, A., "Measure of the ion implantation doping uniformity", Acta Electron., 19 (1), 77-86 (1976), French, PA79-66109.
245. Morozov, N. P. and Tetel'baum, D. I., "Radiation defect formation during ion implantation of semiconductors in the presence of force fields", Phys. Status Solidi A, 51 (2), 629-40 (1979), CA90-161061.
246. Mueller, G., Trapp, M., Schimko, R. and Richter, C. E., "Measurement of range distributions of zinc and nitrogen ions in multiple-layer substrates with the secondary ion microprobe", Phys. Status Solidi A, 51 (1), 87-92 (1979), CA90-144924.

247. Murakami, K., Gamo, K., Namba, S., Kawabe, M. and Aoyagi, Y., "Dynamic behavior of mode-locked Nd:YAG laser annealing in ion-implanted Si, GaAs, and GaP", Appl. Phys. Lett., 35 (8), 628-30 (1979), EEA83-9493.
248. Myers, D. R., Koyama, R. Y. and Phillips, W. E., "An implantation predeposition technique for the introduction of deep-level chemical impurities", Proceedings of the International Conference on Ion Beam Modification of Materials, 1978, p. 439-48.
249. Mylroie, S. W., "Improved range statistics for implanted ions in semiconductors", Thesis, Stanford Univ., Stanford, Calif., 1975, 143 pp., Order No. 76-5775, CA84-158745.
250. Nakamura, K. and Kamoshida, M., "Implanted As redistribution during annealing in oxidizing ambient", J. Electrochem. Soc., 125 (9), 1518-21 (1978), EEA82-12569.
251. Nakatsuka, M., Tanaka, K. and Kikkawa, T., "Effects of thin conductive film mask on ion implantation", J. Electrochem. Soc., 125 (11), 1829-32 (1978).
252. Narayan, J. and White, C. W., "Extent of annealed or melted regions as a function of energy of pulsed laser irradiation", AIP Conf. Proc., no. 50, 446-52 (1978), EEA82-38232.
253. Natsuaki, N., Tamura, M. and Tokuyama, T., "Nonequilibrium solid solutions obtained by heavy ion implantation and laser annealing", J. Appl. Phys., 51 (6), 3373-82 (1980), EEA83-44000.
254. Nishi, H., Inada, T., Sakurai, T., Kaneda, T., Hisatsugu, T. and Furuya, T., "Uniform doping of channeled-ion implantation", J. Appl. Phys., 49 (2), 608-13 (1978), EEA81-28415.
255. Nishi, H., Sato, T. and Furuya, T., "Uniform and sloping profiles obtained by multiple dose-energy implantation", Fujitsu Sci. & Tech. J., 11 (4), 113-22 (1975), EEA79-16604.
256. Oetzmann, H., Feuerstein, A., Grahmann, H. and Kalbitzer, S., "Range parameters of heavy ions in amorphous targets at LSS energies of $0.0006 \leq \epsilon \leq 0.3$", Phys. Lett. A, 55A (3), 170-2 (1975), CA84-155988.
257. Oetzmann, H. and Kalbitzer, S., "^{4}He-stopping power measurements by using ion implantation and backscattering techniques", Radiat. Eff., 47 (1-4), 73-80 (1980), PA83-88233.
258. Ohdomari, I., "Formation of amorphous layer by ion implantation", Oyo Buturi, 46 (8), 793-802 (1977), Japanese, EEA81-14656.
259. Oka, Y., "Ion implantation with controlled non uniformity", J. Vac. Sci. Technol., 17 (4), 817-19 (1980), EEA83-53651.
260. Park, Y. S., "Reaction of semiconductors to ion implantation and electron bombardment", Report AD-A008737/9SL, Aerospace Research Labs., Wright-Patterson AFB, Ohio, 1975, 51 pp.
261. Parry, P. D., "Target heating during ion inplantation", J. Vac. Sci. & Technol., 13 (2), 622-9 (1976), PA79-69537.
262. Parry, P. D., "Localized substrate heating during ion implantation", J. Vac. Sci. & Technol., 15 (1), 111-15 (1978), EEA81-28417.

263. Pham, M. T., "Determination of ion implantation profiles from measurement of the differential capacitance in semiconductor-electrolyte contacts", Phys. Status Solidi A, 37 (2), 439-43 (1976), German, EEA80-619.
264. Plew, L. E., "Automated SIMS concentration profiling of ion implanted dopants in semiconductors", Electrochemical Society Spring Meeting (Extended Abstracts), 1976, p. 521-2, PA79-91566.
265. Prussin, S., "The interaction of oxidation, diffusion, and implanted defect nuclei", Electrochemical Society Fall Meeting, Extended Abstracts, 1979, p. 1227-31.
266. Rabette, P., Deane, A. M., Tench, A. J. and Che, M., "Catalytic activity of ion implanted single crystal supports: A new tool in catalysis studies", Chem. Phys. Lett., 60 (2), 348-52 (1979), PA82-20462.
267. Ratnakumar, K. N., Pease, R. F. W., Bartelink, D. J. and Johnson, N. M., "Scanning electron beam annealing with a modified SEM", J. Vac. Sci. & Technol., 16 (6), 1843-6 (1979), EEA83-34769.
268. Restelli, G. and Cappellani, F., "Ion beam techniques for semiconductor studies", Report EUR-5808EN, Comm. European Communities, Luxembourg, 1978, 169 pp., EEA83-9710.
269. Rimini, E., Bacri, P., Campisano, S. U. and Foti, G., "Ruby laser pulse effects in ion implanted semiconductors", AIP Conf. Proc., no. 50, 259-73 (1978), EEA82-34272.
270. Rodine, E. T., Berliner, R. R. and Shin, B. K. "Ion implantation in compound semiconductors", Report AD-A037958, Syst. Res. Lab. Inc., Dayton, Ohio, 1977, 157 pp. CA87-94357.
271. Ross, R., "Dependence of the etching rate on the dose and the type of implanted ions in profile determination", Zentralinst, Kernforsch., Rossendorf Dresden, ZfK (294), 211-12 (1975), German, CA86-82541.
272. Ross, R., "Measurement of sputtering rates in phosphorus implantation", Zentralinst. Kernforsch., Rossendorf Dresden, ZfK (294), 214-15 (1975), German, CA86-82543.
273. Rozgonyi, G. A. and Seidel, T. E., "Influence of wafer orientation and ion implantation on surface and bulk stacking faults", Electrochemical Society Spring Meeting, 1977, p. 543-5, EEA81-5974.
274. Runge, H., "Distribution of implanted ions under arbitrarily shaped mask edges", Phys. Status Solidi A, 39 (2), 595-9 (1977), EEA80-20920.
275. Rupprecht, H. S. and Schwenker, R. O., "Implanting ions in a semiconductor substrate", Patent Germany 2801271, Publ. August 1978, CA89-139374.
276. Santry, D. C., "Preparation of targets by ion implantation", Report AECL-5503, Chalk River Nucl. Lab., At. Energy Canada Ltd., Chalk River, Ont., 1975, p. 1-13, CA85-84404.

277. Sato, T., Nishi, H. and Furuya, T., "Redistributions of implanted boron and phosphorus sources during thermal diffusion", Fujitsu Sci. Tech. J., 12 (1), 115-24 (1976), CA85-12873.
278. Schaumburg, H., "Doping uniformity and reproducibility of ion implanted wafers", Acta Electron., 19 (2), 141-5 (1976), CA85-115530.
279. Schmitt, A. and Wagner, H., "Wafer cooling during ion implantation", IBM Tech. Disclosure Bull., 22 (10), 4600 (1980).
280. Sealy, B. J., "Laser annealing of ion implanted semiconductors", J. Cryst. Growth, 48 (4), 655-65 (1980), EEA83-29121.
281. Setvak, M., "Use of ion beams in thin film analysis and technology", Jemna Mech. & Opt., 21 (11), 325-8 (1976), EEA80-24875.
282. Shah, P., "Analysis of implanted layer in silicon for semiconductor device processing", 4th Conference on the Scientific and Industrial Applications of Small Accelerators, Proceedings, 1976, p. 411-19, EEA80-36259.
283. Sizmann, R. and Varelas, C., "Surface channeling", Nucl. Instrum. Methods, 132, 633-8 (1976), CA85-12459.
284. Smirnov, L. S. and Kachurin, G. A., "Laser annealing of ion-implanted semiconductors", Proceedings of the International Conference on Ion Beam Modification of Materials, 1978, p. 653-68.
285. Smith, B. J., "Capacitance-voltage profiling techniques", Report AERE-R8862, UKAEA, Harwell, Oxon, England, 1977, 33 pp. EEA81-19430.
286. Speriosu, V. S., Glass, H. L. and Kobayashi, T., "X-Ray determination of strain and damage distributions in ion-implanted layers", Appl. Phys. Lett., 34 (9), 539-42 (1979), PA82-69116.
287. Spitzer, W. G., "Optical studies of ion implanted species in group IV and III-V semiconductors", Report AD-A029553, Electron. Sci. Lab., Univ. South. California, Los Angeles, Calif., 1976, 17 pp., CA86-64419.
288. Srobar, F., "A method of indirect doping of solids", Cesk. Cas. Fyz. Sekce A, 29 (2), 174-5 (1979), Czech., EEA83-9692.
289. Stein, H. J., "Atomic displacement effects in neutron transmutation doping", Neutron Transmutation Doping in Semiconductors, 1978, p. 229-48, EEA83-9730.
290. Stephens, K. G. and Wilson, I. H., "Properties and applications of ion-implanted films", Thin Solid Films, 50, 325-47 (1978), EEA81-42359.
291. Syhre, H. and Deutscher, M., "Ion implantation", Patent East Germany 131103, Publ. May 1978, CA90-79913.
292. Tan, P. M., "Determination of ion implantation profiles in the measurement of the differential capacitance of a semiconductor electrolyte contact", Phys. Status Solidi A, 37 (2), 439-44 (1976), German, CA86-11198.

293. Tebo, A. R., "Laser annealing - and beyond", Electro-Opt. Syst. Des., 11 (12), 29-34 (1979), EEA83-23883.
294. Thomas, G. J. and Bauer, W., "In-situ observations of ion implanted surfaces", J. Nucl. Mater., 63, 280-4 (1976), PA80-35598.
295. Thompson, D. A. and Walker, R. S., "Channeling measurements of damage in ion bombarded semiconductors at 50K", Radiat. Eff., 30 (1), 37-46 (1976), PA79-79401.
296. Titov, V. V., "Control of impurity profile by ion implantation", Sov. Phys.-Tech. Phys., 23 (11), 1378-80 (1978), PA82-82258.
297. Trejbalova, V., "Ion implantation computer modelling", Scr. Fac. Sci. Nat. Univ. Purkynianae Brunensis Phys., 6 (1), 53-8 (1976), PA80-43235.
298. van der Weg, W. and Tamminga, Y., "High energy ion scattering as a tool for studying ion implantation", Acta Electron., 19 (1), 41-52 (1976), PA79-66106.
299. Vitali, G., Bertolotti, M. and Stagni, I., "Laser induced structure changes in implanted semiconductors related to pulse shape", AIP Conf. Proc., no. 50, 111-15 (1978), EEA82-34269.
300. Wada, Y., Usui, H. and Ashikawa, M., "Substrate temperature measurement during ion implantation", Jap. J. Appl. Phys., 14 (9), 1351-6 (1975), EEA79-832.
301. Walker, R. S. and Thompson, D. A., "Measurement of damage distributions in ion bombarded Si, GaP and GaAs at 50K", Nucl. Instrum. & Methods, 135 (3), 489-95 (1976), EEA79-37459.
302. White, C. W., Christie, W. H., Pronko, P. P., Appleton, B. R., Wilson, S. R., Young, R. T., Wang, J. C., Wood, R. F. and Narayan, J., "Dopant profile changes induced by pulsed laser annealing", Radiat. Eff., 47 (1-4), 37-40 (1980), EEA83-4406.
303. White, C. W., Narayan, J., Appleton, B. R. and Wilson, S. R., "Impurity segregation by pulsed laser irradiation", J. Appl. Phys., 50 (4), 2967-9 (1979), EEA82-38254.
304. White, C. W., Narayan, J. and Young, R. T., "Laser annealing of ion-implanted semiconductors", Science, 204 (4392), 461-8 (1979), EEA82-38256.
305. Wilson, R. G., Dunlap, H. L., Jamba, D. M. and Myers, D. R., "Semiconductor measurement technology: angular sensitivity of controlled implanted doping profiles", Report NBS-SP-400-49, Electron Devices Div., Natl. Eng. Lab., Washington, DC, 1978, 64 pp., CA91-67260.
306. Wolf, G. K., "Chemical and electrochemical aspects of ion implantation", Radiat. Eff., 48 (1-4), 237-46 (1980), PA83-100271.
307. Wu, C. P., Douglas, E. C. and Mueller, C. W., "Limitations of the CV technique for ion-implanted profiles", IEEE Trans. Electron Devices, ED-22 (6), 319-29 (1975), PA78-47424.
308. Wu, C. P., Douglas, E. C. and Mueller, C. W., "Obtaining the distribution of profile of electrically active ions implanted in a semiconductor", Patent USA 3976377, Publ. August 1976, CA85-136088.

309. Wu, C. P., Douglas, E. C., Mueller, C. W. and Williams, R., "Techniques for lapping and staining ion-implanted layers", J. Electrochem. Soc., 126 (11), 1982-8 (1979).
310. Yoshikawa, A., Tamama, T. and Mizushima, Y., "Properties of Se-Ge inorganic photoresist as an ion-implantation mask", Electron Lett., 15 (16), 499-501 (1979), EEA82-47075.
311. Zaremba, A. J., Marcyk, G. T. and Streetman, B. G., "Optimal summation of gaussians for ion implantation profile control", IEEE Trans. Electron Devices, ED-24 (2), 163-5 (1977).
312. Ziegler, J. F. and Chu, W. K., "Tables of stopping cross sections for ^{4}He ions in solids", J. Appl. Phys., 47 (5), 2239-40 (1976), PA79-68582.
313. Zuleeg, R. and Lehovec, K., "Neutron degradation of ion-implanted and uniformly doped enhancement mode", IEEE Trans. Nucl. Sci., NS-25 (6), 1444-9 (1978).

2. Group IV Semiconductors

A. Silicon

a. Amorphous

314. Baranova, E. K., Gusev, V. M. and Khaibullin, I. B., "Influence of irradiation temperature on the amorphization of silicon by ion bombardment", Sov. Phys.-Semicond., 9 (5), 630-1 (1975), PA79-13540.
315. Beyer, W., Barna, A. and Wagner, H., "Highly doped evaporated amorphous silicon by alkali implantation", Appl. Phys. Lett., 35 (7), 539-41 (1979), EEA83-731.
316. Beyer, W., Fischer, R. and Wagner, H., "Lithium doping of amorphous silicon", J. Electron Mater., 8 (2), 127-37 (1979).
317. Beyer, W., Stritzker, B. and Wagner, H., "Influence of alkali and halogen implantation on electrical properties of amorphous silicon", J. Non Cryst. Solids, 35+36, Pt. 1, 321-6 (1980).
318. Csepregi, L., Mayer, J. W. and Sigmon, T. W., "Channeling effect measurements of the recrystallization of amorphous Si layers on crystal Si", Phys. Lett. A, 54A (2), 157-8 (1975), PA79-1907.
319. Csepregi, L., Mayer, J. W. and Sigmon, T. W., "Regrowth behavior of ion-implanted amorphous layers on (111) silicon", Appl. Phys. Lett., 29 (2), 92-3 (1976), EEA79-37410.
320. Dennis, J. R., "A study of the amorphization of silicon by ion implantation", Thesis, University of Missouri, Rolla, MO, 1976, 125 pp., Order No. 76-26857.
321. Dennis, J. R. and Hale, E. B., "Energy dependence of amorphizing implant dose in silicon", Appl. Phys. Lett., 29 (9), 523-4 (1976), CA85-185618.
322. Dennis, J. R. and Hale, E. B., "Amorphization of silicon by ion implantation: homogeneous or heterogeneous nucleation", Radiat. Eff., 30 (4), 219-25 (1976), PA80-11578.

323. Dennis, J. R. and Hale, E. B., "Crystalline to amorphous transformation in ion-implanted silicon: a composite model", J. Appl. Phys., 49 (3), 1119-27 (1978).
324. Dvurechenskii, A. V. and Smirnov, L. S., "Localized states and conductivity in silicon amorphized by ion implantation", Phys. Status Solidi A, 56 (2), 647-54 (1979), PA83-39529.
325. Foti, G., Rimini, E., Vitali, G. and Bertolotti, M., "Amorphous polycrystal transition induced by laser pulse in self-ion implanted silicon", Appl. Phys., 14 (2), 189-91 (1977), EEA80-41216.
326. Gerasimenko, N. N., Gudaev, G. A., Dvurechenskii, A. V., Smirnov, L. S. and Sokolov, S. A., "Amorphization of silicon due to irradiation with heavy ions", Sov. Phys.-Semicond., 10 (7), 735-7 (1976), EEA80-6485.
327. Gibson, J. B., "High resolution microscopy of ion-implanted amorphous silicon", Electron Microscopy and Analysis Symposium, 1979, p. 149-52, Published by Institute of Phys., 1980, EEA83-33770.
328. Gotz, G., Karthe, W., Schnable, B. and Sobolov, N., "ESR line shape study of amorphous centres in ion implanted silicon", Phys. Status Solidi, A, 50 (2), K209-12 (1978), PA82-28587.
329. Hale, E. B., "Mechanisms for amorphous state formation in ion implanted silicon", Report AD-A055480, Missouri Univ., Rolla, MO, 1978, 8 pp.
330. Kalbitzer, S., Muller, G., Le Comber, P. G. and Spear, W. E., "The effects of ion implantation on the electrical properties of amorphous silicon", Philos. Mag. B, 41 (4), 439-56 (1980), EEA83-39759.
331. Khokhlov, A. F., "Ion implantation in amorphous silicon", Sov. Phys.-Tech. Phys., 49 (11), 1373-4 (1979), EEA83-53665.
332. Khokhlov, A. F., Mashin, A. I. and Polyakov, S. M., "Magnetic ordering in silicon made amorphous by ion implantation", Sov. Phys.-Semicond., 12 (5), 549-51 (1978), PA82-24186.
333. Klimenko, A. G., Klimenko, E. A. and Donin, V. I., "Use of argon laser radiation in restoration of single-crystal state of ion-implantation-amorphized silicon surface", Sov. J. Quantum Electron., 5 (10), 1289-91 (1975), PA79-54475.
334. Le Comber, P. G., Spear, W. E., Mueller, G. and Kalbitzer, S., "Electrical and photoconductive properties of ion implanted amorphous silicon", J. Non Cryst. Solids, 35+36, Pt. 1, 327-32 (1980).
335. Matsumura, H., Kuzuta, N., Ishiwara, H. and Furukawa, S., "Resistivity control of sputtered amorphous silicon by ion implantation", Radiat. Eff., 48 (1-4), 133-8 (1980), EEA83-53661.
336. Miyao, M., Tamura, M. and Tokuyama, T., "Selective annealing of ion-implanted amorphous layers by Nd^{3+}-YAG laser irradiation", Appl. Phys. Lett., 33 (9), 828-30 (1978), EEA82-8260.

337. Muller, G. and Kalbitzer, S., "Electrical conductivity of ion implanted amorphous silicon", 4th International Conference on the Physics of Non-Crystalline Solids, 1977, p. 278-83, EEA81-14521.
338. Muller, G. and Kalbitzer, S., "The crystalline-to-amorphous transition in ion-bombarded silicon", Philos. Mag. B, 41 (3), 307-25 (1980), EEA83-34654.
339. Muller, G., Kalbitzer, S., Spear, W. E. and Le Comber, P. G., "Doping of amorphous silicon by ion implantation", Amorphous Liq. Semicond., 7th International Conference, Proceedings, 1977, p. 442-6, CA89-98604.
340. Muller, J. C., Grob, J. J., Grob, A., Stuck, R. and Siffert, P., "Laser annealing of silicon layers amorphized by molecular ions", Radiat. Eff., 48 (1-4), 115-20 (1980), EEA83-53660.
341. Ohmura, Y., Inoue, T. and Yamamoto, Y., "Spatially varied activation of ion-implanted As during the regrowth of amorphous layers in Si", J. Appl. Phys., 49 (6), 3597-9 (1978), EEA81-38125.
342. Peercy, P. S. and Stein, H. J., "Laser annealing of hydrogen implanted amorphous silicon", AIP Conf. Proc., no. 50, 331-6 (1978), EEA82-34280.
343. Revesz, P., Wittmer, M., Roth, J. and Mayer, J. W., "Epitaxial regrowth of Ar-implanted amorphous silicon", J. Appl. Phys., 49 (10), 5199-206 (1978), EEA82-8109.
344. Spear, W. E., Le Comber, P. G., Kalbitzer, S. and Muller, G., "Doping of amorphous silicon by alkali-ion implantations", Philos. Mag. B, 39 (2), 159-65 (1979), EEA82-30288.
345. Spitzer, W. G., Ko, J. S. and Waddell, C. N., "Plasma region in high-fluence implants of phosphorus in amorphized silicon", J. Appl. Phys., 50 (5), 3775-7 (1979), EEA82-42456.
346. Spitzer, W. G., Waddell, C. N., Narayanan, G. H., Fredrickson, J. E. and Prussin, S., "Free-carrier plasma effects in ion-implanted amorphous layers of silicon", Appl. Phys. Lett., 30 (12), 623-6 (1977), EEA80-36031.
347. Tamminga, Y. and Josquin, W. J. M. J., "The regrowth of amorphous layers created by high-dose antimony implantation in (100) silicon", Appl. Phys. Lett., 32 (1), 13-15 (1978), EEA81-19414.
348. Tsai, M. Y., Day, D. S., Streetman, B. G., Williams, P., and Evans, C. A., Jr., "Recrystallization of implanted amorphous silicon layers. II. Migration of fluorine in BF_2^+-implanted silicon", J. Appl. Phys., 50 (1), 188-92 (1979), EEA82-30374.
349. Tsai, M. Y. and Streetman, B. G., "Recrystallization of implanted amorphous silicon layers. I. Electrical properties of silicon implanted with BF_2^+ or $Si^+ + B^+$", J. Appl. Phys., 50 (1), 183-7 (1979), EEA82-30373.
350. Wittmer, M., Roth, J., Revesz, P. and Mayer, J. W., "Epitaxial regrowth of Ne- and Kr-implanted amorphous silicon", J. Appl. Phys., 49 (10), 5207-12 (1978), EEA82-8110.

351. Yasaitis, J. A., "Ion implantation of neon in silicon for planar amorphous isolation", Electron. Lett., 14 (15), 460-2 (1978), EEA81-42437.

b. Annealing

352. Antonenko, A. K., Gerasimenko, N. N., Dvurechenskii, A. V., Smirnov, L. S. and Tseitlin, G. M., "Distribution of an implanted impurity in silicon after laser annealing", Sov. Phys. -Semicond., 10 (1), 81-2 (1976), EEA80-3467.
353. Appleton, B. R., White, C. W., Larson, B. C., Wilson, S. R. and Narayan, J., "Laser processing of ion implanted silicon", IEEE Trans. Nucl. Sci., NS-26 (1), pt. 2, 1686-92 (1979), EEA82-26633.
354. Auston, D. H., Golovchenko, J. A., Smith, P. R., Surko, C. M. and Venkatesan, T. N. C., "CW argon laser annealing of ion-implanted silicon", Appl. Phys. Lett., 33 (6), 539-41 (1978), EEA82-794.
355. Auston, D. H., Surko, C. M., Venkatesan, T. N. C., Slusher, R. E. and Golovchenko, J. A., "Time-resolved reflectivity of ion-implanted silicon during laser annealing", Appl. Phys. Lett., 33 (5), 437-40 (1978), EEA81-49610.
356. Battaglin, G., Della Mea, G., Drigo, A. V., Foti, G., Bentini, G. G. and Servidori, M., "Two-stage laser annealing of lattice disorder in phosphorus implanted silicon", Phys. Status Solidi A, 49 (1), 347-52 (1978), EEA82-810.
357. Beanland, D. G. and Williams, J. S., "The damage dependence of the epitaxial regrowth rate during the annealing of amorphous silicon formed by ion implantation", Radiat. Eff., 36 (1-2), 15-21 (1978), EEA81-28270.
358. Bentini, G. G., Galloni, R. and Nipoti, R., "Multiscanning electron beam annealing of phosphorus-implanted silicon", Appl. Phys. Lett., 36 (8), 661-3 (1980), EEA83-34606.
359. Bocanek, L., "The influence of annealing on the distribution and mobility of carriers in phosphorus implanted silicon", Scr. Fac. Sci. Nat. Univ. Purkynianae Brunensis Phys., 8 (1), 1-10 (1978), EEA81-42235.
360. Bomke, H. A., Berkowitz, H. L., Harmatz, M., Kronenberg, S. and Lux, R., "Annealing of ion-implanted silicon by an incoherent light pulse", Appl. Phys. Lett., 33 (11), 955-7 (1978), EEA82-12436.
361. Bourguet, P. and Dupart, J. M., "Nitrogen implantation at high doses in monocrystalline silicon: doses and annealing conditions for obtaining a buried homogeneous insulating layer", Rev. Phys. Appl., 15 (3), 647-52 (1980), EEA83-29141.
362. Campisano, S. U., Foti, G. and Servidori, M., "Laser irradiation of furnace preannealed ion implanted silicon", Appl. Phys. Lett., 36 (4), 279-81 (1980), PA83-65015.

363. Celler, G. K., "CW infrared laser annealing of ion-implanted silicon", J. Appl. Phys., 50 (11), pt. 1, 7264-6 (1979), EEA83-14292.
364. Celler, G. K., Borutta, R., Brown, W. L., Poate, J. M., Rozgonyi, G. A. and Sheng, T. T., "CO_2 laser annealing of ion-implanted silicon", AIP Conf. Proc., no. 50, 381-6 (1978), EEA82-38222.
365. Celler, G. K., Brown, W. L., Kimerling, L. C., Leamy, H. J., Miller, G. L., Poate, J. M., Rozgonyi, G. A. and Sheng, T. T., "Characterisation of ion-implanted silicon annealed by laser irradiation", International Conference on Defects and Radiation Effects in Semiconductors, 1978, p. 460-7, EEA83-14176.
366. Celler, G. K., Poate, J. M. and Kinerling, L. C., "Spatially controlled crystal regrowth of ion-implanted silicon by laser irradiation", Appl. Phys. Lett., 32 (8), 464-6 (1978), EEA81-33096.
367. Christodoulides, C. E., "Impurity redistribution in Pb implanted and annealed silicon", Phys. Lett. Vol. A, 67A (1), 83-6 (1978), EEA81-42354.
368. Chu, W. K., "Stability study of laser irradiation of silicon diffused with arsenic", Appl. Phys. Lett., 36 (4), 273-5 (1980).
369. Chu, W. K., Masters, B. J. and Riseman, J., "Contact formation by laser annealing of implanted silicon", IBM Tech. Disclosure Bull., 22 (12), 5335 (1980), EEA83-53672.
370. Correra, L. and Pedulli, L., "Incoherent-light-flash annealing of phosphorus-implanted silicon", Appl. Phys. Lett., 37 (1), 55-7 (1980), PA83-97504.
371. Cullis, A. G., Poate, J. M. and Celler, G. K., "Redistribution and enhanced solubility of transition metal implants in silicon following laser irradiation", AIP Conf. Proc., no. 50, 311-16 (1978), EEA82-34276.
372. Cullis, A. G., Seidel, T. E. and Meek, R. L., "Comparative study of annealed neon-, argon-, and krypton-ion implantation damage in silicon", J. Appl. Phys., 49 (10), 5188-98 (1978), EEA82-8263.
373. Cullis, A. G., Webber, H. C., Poate, J. M. and Simons, A. L., "Segregation and increased dopant solubility in Pt-implanted and laser-annealed Si layers", Appl. Phys. Lett., 36 (4), 320-23 (1980).
374. Damgaard, S., Oron, M., Petersen, J. W., Petrikin, Y. V. and Weyer, G., "Laser annealing studies on ion implanted iron in silicon", Phys. Status Solidi A, 59 (1), 63-7 (1980), PA83-79827.
375. Deval, S. and Nuttall, R., "Enhanced anneal behaviour of boron implanted silicon", Solid-State Electron., 20 (4), 381-2 (1977), EEA80-20922.
376. Dvurechensky, A. V. and Ryazantsev, I. A., "On the nature of the defect reverse annealing in ion-implanted silicon", Radiat. Eff., 46 (3-4), 129-32 (1980), PA83-88875.

377. Eggermont, G. E. J., Tamminga, Y. and Hofker, W. K., "Elemental- and dose dependent threshold for Nd-YAG laser induced recrystallization of silicon", AIP Conf. Proc., no. 50, 321-4 (1978), EEA82-34278.
378. Engstrom, H. and Bates, J. B., "Raman scattering from boron-implanted laser-annealed silicon", J. Appl. Phys., 50 (4), 2921-5 (1979), EEA82-38047.
379. Foti, G., Rimini, E. and Campisano, S. U., "Laser annealing of Pb-implanted silicon", Phys. Status Solidi A, 47 (2), 533-8 (1978), EEA81-42355.
380. Gat, A., Gerzberg, L., Gibbons, J. F., Lietoila, A., Johnson, N. M., Magee, T. J., Peng, J., Deline, V., Williams, P. and Evans, C. A., Jr., "Annealing of ion-implanted Si using a scanned CW laser system", Radiat. Eff., 48 (1-4), 195-202 (1980), EEA83-48713.
381. Gat, A., Gibbons, J. F., Magee, T. J., et al., "Physical and electrical properties of laser-annealed ion-implanted silicon", Appl. Phys. Lett., 32 (5), 276-8 (1978), EEA81-28257.
382. Gat, A., Gibbons, J. F., Magee, T. J., et al., "Use of a scanning CW Kr laser to obtain diffusion-free annealing of B-implanted silicon", Appl. Phys. Lett., 33 (5), 389-91 (1978), EEA81-49712.
383. Gat, A., Lietoila, A. and Gibbons, J. F., "A study of the mechan ism of CW laser annealing of arsenic-implanted silicon", J. Appl. Phys., 50 (4), 2926-9 (1979), EEA82-38048.
384. Gat, A., Magee, T. J., Peng, J., Deline, V. R. and Evans, C. A., Jr., "CW laser annealing of boron and arsenic-implanted silicon; electrical properties, crystalline structure and limitations", Solid State Technol., 22 (11), 59-68 (1979), EEA83-19004.
385. Gavrilenko, V. I., Kalandadze, T. M., Karplyuk, A. I., Litovchenko, V. G. and Torchun, N. M., "Investigating the processes of annealing ion-implanted silicon structures", Poluprovodn. Tekh. & Mikro-elektron., no. 30, 45-55 (1979), Russian, EEA83-23921.
386. Geiler, H. D. and Weibflog, R., "Laser-induced annealing effects of radiation damage in silicon", Proceedings of the International Conference on Ion Beam Modification of Materials, 1978, p. 699-706.
387. Gibbons, J. F., Gat., A., Gerzberg, L., Lietoila, A., Regolini, J. L., Sigmon, T. W., Pease, R. F. W., Magee, T. J., Peng, J., Hong, J., Deline, V., Katz, W., Williams, P. and Evans, C. A., Jr., "Annealing of ion-implanted Si using scanned laser and electron beams", AIP Conf. Proc., no. 50, 365-80 (1978), EEA82-38221.
388. Giergiel, J. and Wroblewska, B., "Redistribution of implanted impurities in silicon during annealing under oxidation conditions", Pr. Nauk. Inst. Technol. Elektron. Politech. Wroclaw., 21, 67-73 (1978), CA90-161037.

389. Grob, J. J., Siffert, P., Prisslinger, R. and Kalbitzer, S., "Damage profiles and annealing of Si(B) implants", Applications of Ion Beams to Materials, 1975, p. 24-30, Publ. 1976, PA79-69553.
390. Guldberg, J., "Electron trap annealing in neutron transmutation doped silicon", Appl. Phys. Lett., 31 (9), 578-9 (1977), CA88-14999.
391. Heinig, K. H., Klabes, R. and Woittennek, H., "A description of the laser-induced annealing and diffusion behaviour of im-implanted silicon crystals", Radiat. Eff., 47 (1-4), 157-61 (1980), EEA83-44016.
392. Hess, L. D., Forber, R. A., Kokorowski, S. A. and Olson, G. L., "CW laser annealing of ion-implanted silicon", Proc. Soc. Photo-Opt. Instrum. Eng., 198, 31-4 (1979), EEA83-48711.
393. Hill, C., "Boundary effects in laser annealing silicon device structures", AIP Conf. Proc., no. 50, 419-24 (1978), EEA82-38227.
394. Hofker, W. K., Eggermont, G. E. J., Tamminga, Y. and Oosthoek, D. P., "The removal of implantation damage in silicon by laser irradiation and its thermal stability", AIP Conf. Proc., no. 50, 425-8 (1978), EEA82-38228.
395. Hofker, W. K., Oosthoek, D. P., Eggermont, G. E. J., Tamminga, Y. and Stacy, W. T., "Laser irradiation of silicon containing misfit dislocations", Appl. Phys. Lett., 34 (10), 690-2 (1979), EEA82-38213.
396. Inada, T., Sugiyama, T., Okano, N. and Ishikawa, Y., "Pulsed electron-beam annealing of phosphorus-implanted silicon", Electron. Lett., 16 (2), 54-5 (1980), EEA83-14283.
397. Itoh, N., Nakau, T., Morikawa, Y. and Nagami, K., "Annealing behaviour of stress in Sb-implanted Si", Jpn. J. Appl. Phys., 17 (6), 1003-8 (1978), EEA81-33101.
398. Itoh, T. and Shinada, K., "Annealing behavior of boron and oxygen implanted silicon", Jap. J. Appl. Phys., 14 (10), 1627-8 (1975), PA79-6408.
399. Johansen, A., Svenningsen, B., Chadderton, L. T. and Whitton, J. L., "Annealing of damage caused by implantation of group IB elements into silicon", Applications of Ion Beams to Materials, 1975, p. 267-72, Publ. 1976, PA79-68574.
400. Johnson, N. M., Gold, R. B., Lietoila, A. and Gibbons, J. F., "Deep levels in ion-implanted, CW laser-annealed silicon", AIP Conf. Proc., no. 50, 550-5 (1978), EEA82-34157.
401. Kachurin, G. A. and Nidaev, E. V., "Diffusion of impurities as a result of laser annealing of implanted layers", Sov. Phys.-Semicond., 11 (3), 350-2 (1977), EEA81-1202.
402. Kannan, V. C. and Casey, D. D., "Two-step annealing of arsenic-implanted (111) silicon", Appl. Phys. Lett., 31 (11), 721-2 (1976), EEA81-10261.

403. Kannan, V. C. and Casey, D. D., "Dose dependence of residual defects in single and two-step annealed arsenic implanted (111) silicon", Electrochemical Society Spring Meeting, Extended Abstracts, 1977, p. 207-8, EEA81-5928.

404. Kaufmann, E. N., Kalish, R., Naumann, R. A. and Lis, S., "Annealing behaviour of In implanted in Si studied by perturbed angular correlation", J. Appl. Phys., 48 (8), 3332-6 (1977), EEA81-1194.

405. Kaufmann, U. and Mitlchner, H., "Annealing behavior of neutron-transmutation doped silicon - as studied by EPR", J. Appl. Phys., 50 (5), 3258-60 (1979), EEA82-42453.

406. Kennedy, E. F., Lau, S. S., Golecki, I., Mayer, J. W., Tseng, W., Minnucci, J. A. and Kirkpatrick, A. R., "Pulsed electron beam annealing of ion implanted Si layers", Radiat. Eff. Lett., 43 (1), 31-6 (1979), EEA82-26645.

407. Khaibullin, I. B., Shtyrkov, E. I., Zaripov, M. M. , Bayazitov, R. M. and Galjautdinov, M. F., "Some features of laser annealing of implanted silicon layers", Radiat. Eff., 36 (3-4), 225-33 (1978), EEA81-38127.

408. Khaibullin, I. B., Shtyrkov, E. I., Zaripov, M. M., Galyautdinov, M. F. and Zakirov, G. G., "Utilization coefficient of implanted impurities in silicon layers subjected to subsequent laser annealing", Sov. Phys.-Semicond., 11 (2), 190-2 (1977), EEA80-36240.

409. Kool, W. H., Roosendaal, H. E., Wiggers, L. W. and Saris, F. W., "Production and beam annealing of damage in argon implanted silicon. II", Radiat. Eff., 36 (1-2), 41-8 (1978), EEA81-28421.

410. Krynicki, J. and Bourgoin, J. C., "Effect of ionisation on defect annealing in phosphorus-implanted silicon", International Conference on Defects and Radiation Effects in Semiconductors, 1978, p. 482-6, EEA83-14177.

411. Krynicki, J., Suski, J., Ugniewski, S., Grotzschel, R., Klabes, R., Kreissig, U. and Rudiger, J., "Laser annealing of arsenic implanted silicon", Phys. Lett. Vol. A, 61A (3), 181-2 (1977), EEA80-28829.

412. Kurnaev, S. and Uggerhøj, E., "Laser annealing of indium-implanted silicon", Radiat. Eff., 40 (1-2), 91-4 (1979).

413. Kutukova, O. G. and Strel'tsov, L. N., "Laser annealing of implanted silicon", Sov. Phys.-Semicond., 10 (3), 265-7 (1976), EEA79-46071.

414. Leamy, H. J., Rozgonyi, G. A., Sheng, T. T. and Celler, G. K., "Periodic regrowth phenomena produced by laser annealing of ion-implanted silicon", Appl. Phys. Lett., 32 (9), 535-7 (1978), EEA81-38000.

415. Lietoila, A. and Gibbons, J. F., "The effect of free-carrier absorption on the annealing of ion-implanted silicon by pulsed lasers", Appl. Phys. Lett., 34 (5), 332-4 (1979), CA90-161064.

416. Lietoila, A., Gibbons, J. F., Magee, T. J., Peng, J. and Hong, J. D., "Solid solubility of As in Si as determined by ion implantation and CW laser annealing", Appl. Phys. Lett., 35 (7), 532-4 (1979), EEA83-614.
417. Little, R. G., Greenwald, A. C. and Minnucci, J. A., "Pulsed electron beams for annealing of ion-implanted silicon", IEEE Trans. Nucl. Sci., NS-26 (1), pt. 2, 1683-5 (1979), EEA82-26632.
418. Liu, Y. S. and Wang, K. L., "A transient optical reflectivity study of laser annealing of ion-implanted silicon: Thresholds and kinetics", Appl. Phys. Lett., 34 (6), 363-5 (1979), EEA82-30236.
419. Luk'yanchuk, T. I., Maksimov, S. K. and Piskunov, D. I., "Defects originating during post-implantation annealing of silicon", In: Svoistva Legir. Poluprovodn., Zemskov, V. S., (Ed.), Moscow, USSR. p. 165-70 (1977), CA89-155706.
420. Mader, S. and Michel, A. E., "Dislocation reactions in arsenic-implanted and annealed silicon", Phys. Status Solidi A, 33 (2), 793-805 (1976), PA79-44915.
421. McMahon, R. A. and Ahmed, H., "Electron-beam annealing of ion-implanted silicon", Electron. Lett., 15 (2), 45-7 (1979), EEA82-17400.
422. Merli, P. G. and Zignani, F., "Self annealing of ion implanted silicon suggestion for an experiment", Radiat. Eff. Lett., 57 (1-2), 59-62 (1980), EEA83-53040.
423. Mitsuishi, T., Sasaki, Y. and Asami, H., "Annealing behavior of phosphorus implanted silicon crystals", Jap. J. Appl. Phys., 16 (2), 367-8 (1977), EEA80-20695.
424. Miyao, M., Natsuaki, N., Yoshihiro, N., Tamura, M. and Tokuyama, T., "Low temperature annealing characteristics of phosphorus-implanted silicon", Jap. J. Appl. Phys., 15 (suppl. 15-1), 57-62 (1976), EEA80-6583.
425. Miyao, M., Ohyu, K. and Tokuyama, T., "Annealing of phosphorus-ion-implanted silicon using a CO_2 laser", Appl. Phys. Lett., 35 (3), 227-9 (1979), EEA83-4241.
426. Miyao, M., Tamura, H., Ohyu, K. and Tokuyama, T., "Annealing behavior of phosphorus implanted silicon irradiated by several lasers of different wavelength", AIP Conf. Proc., no. 50, 325-30 (1978), EEA82-34279.
427. Morhange, J. F., Kanellis, G., Balkanski, M., Peray, J. F., Icole, J. and Croset, M., "Raman spectroscopy of pulsed-laser annealed ion implanted silicon", AIP Conf. Proc., no. 50, 429-33 (1978), EEA82-38229.
428. Muller, J. C., Grob, A., Grob, J. J., Stuck, R. and Siffert, P., "Laser-beam annealing of heavily damaged implanted layers on silicon", Appl. Phys. Lett., 33 (4), 287-9 (1978), CA89-139267.

429. Muller, J. C., Grob, A., Grob., J. J., Stuck, R. and Siffert, P., "Laser annealing of silicon layers amorphized by molecular ions", Proceedings of the International Conference on Ion Beam Modification of Materials, 1978, p. 761-76.
430. Murakami, K., Gamo, K., Kawabe, M., Namba, S. and Aoyagi, Y., "Dynamic behavior of 30-ps pulsed-laser annealing in ion-implanted Si", Jpn. J. Appl. Phys., 18 (12), 2311-12 (1979), EEA83-14169.
431. Murakami, K., Gamo, K., Namba, S., Kawabe, M. and Aoyagi, Y., "Dynamic behaviour of pulsed-laser annealing in ion-implanted silicon. Measurement of the time dependent optical reflectance", Phys. Lett. Vol. A, 70A (4), 332-4 (1979), EEA82-60393.
432. Mustafin, T. N., Kachurin, G. A., Popov, V. P., Pridachin, N. B., Seryapin, V. G. and Smirnov, L. S., "Diffusion of zinc during laser annealing of implanted silicon layers", Sov. Phys. -Semicond., 12 (7), 776-9 (1978), EEA82-17409.
433. Narayan, J., Larson, B. C. and Christie, W. H., "Effect of thermal annealing in boron implanted, laser annealed silicon", AIP Conf. Proc., no. 50, 440-5 (1978), EEA82-38231.
434. Narayan, J., Young, R. T. and White, C. W., "A comparative study of laser and thermal annealing of boron-implanted silicon", J. Appl. Phys., 49 (7), 3912-17 (1978), EEA81-42347.
435. Natsuaki, N., Tamura, M. and Tokuyama, T., "Nonequilibrium solid solutions obtained by heavy ion implantation and laser annealing", J. Appl. Phys., 51 (6), 3373-82 (1980), EEA83-44000.
436. Niu, H., Matsuda, T. and Takai, M., "Study of anneal behaviors of N_2^+-implanted p-Si by means of a lateral photovoltaic method" Jpn. J. Appl. Phys., 17 (11), 1983-92 (1978), EEA82-8265.
437. Nojima, S., Yamazaki, H., Harada, H. and Fujimoto, M., "Annealing characteristics of arsenic-implanted silicon", Jap. J. Appl. Phys., 16 (1), 193-4 (1977), EEA80-9406.
438. Orel, V., "An annealing study of silicon implanted by P^+ ions", Scr. Fac. Sci. Nat. Univ. Purkynianae Brunensis Phys., 9 (1), 29-40 (1979), EEA83-19006.
439. Paine, B. M., Williams, J. S., Austin, M. W. and Celler, G. K., "Comparison of laser annealing (pulsed and continuous) and furnace annealing of low-solid solubility, Pb-implanted Si", AIP Conf. Proc., no. 50, 387-92 (1978), EEA82-38223.
440. Poate, J. M., Bean, J. C., Brown, W. L., et al., "Liquid and solid phase regrowth of Si by laser irradiation and thermally assisted flash annealing", Proceedings of the International Conference on Ion Beam Modification of Materials, 1978, p. 787-806.
441. Prussin, S. and von der Ohe, W., "Laser annealing of low-fluence ion-implanted silicon", J. Appl. Phys., 51 (7), 3853-9 (1980), PA83-97509.
442. Ratnakumar, K. N., Pease, R. F. W., Bartelink, D. J., Johnson, N. M. and Meindl, J. D., "High-resolution scanning electron-beam annealing of ion-implanted silicon", Appl. Phys. Lett., 35 (6), 463-6 (1979), EEA83-4427.

443. Regolini, J. L., Sigmon, T. W. and Gibbons, J. F., "Metastable ^{75}As concentrations formed by scanned CW e-beam annealing of ^{75}As-implanted silicon", Appl. Phys. Lett., 35 (2), 114-16 (1979), EEA82-47074.
444. Regolini, J. L., Sigmon, T. W., Gibbons, J. F., Magee, T. J. and Peng, J., "Recrystallization of ^{75}As implanted silicon by a free running ruby laser", AIP Conf. Proc., no. 50, 393-8 (1978), EEA82-38036.
445. Rimini, E., Baeri, P. and Foti, G., "Laser pulse energy dependence of annealing in ion implanted Si and GaAs semiconductors", Phys. Lett. Vol. A, 65A (2), 153-5 (1978), PA81-34528.
446. Rimini, E., Chu, W. K., Baglin, J. E. E., Tan, T. Y. and Hodgson, R. T., "Laser annealing of silicon implanted with both argon and arsenic", Appl. Phys. Lett., 37 (1), 81-3 (1980), PA83-97506.
447. Romanov, S. T., Kachurin, G. A. and Smirnov, L. S., "Crystallization of ion implanted layers by the nanosecond laser pulses", Radiat. Eff., 48 (1-4), 191-4 (1980), EEA83-53039.
448. Roosendaal, H. E., Kool, W. H. and Saris, F. W., "Production and beam annealing of damage in carbon implanted Si. I", Radiat. Eff., 36 (1-2), 35-40 (1978), EEA81-28420.
449. Rzewuski, H., Suski, J. and Krynicki, J., "Ionization assisted annealing of boron implanted silicon", Radiat. Eff., 39 (2), 123-5 (1978), EEA82-12440.
450. Rzewuski, H., Suski, J., Werner, Z., Csepregi, C. and Gyulai, J., "Electron irradiation assisted annealing of boron and phosphorus implanted silicon layers", Radiat. Eff., 29 (3), 137-41 (1976), PA79-76647.
451. Sasaki, Y., Mitsuishi, T., Kogure, S. and Tanaka, H., "Annealing characteristics of proton-implanted p-type silicon", Jap. J. Appl. Phys., 16 (2), 377-8 (1977), EEA80-42217.
452. Schmitt, A. and Schorer, G., "Damage anneal of antimony/phosphorus double implants in silicon", Appl. Phys., 22 (2), 137-43 (1980), EEA83-43998.
453. Scovell, P. D. and Young, J. M., "Low temperature thermal annealing of arsenic implanted silicon", Electron. Lett., 16 (16), 614-15 (1980), EEA83-48696.
454. Seshan, K. and EerNisse, E. P., "Role of stresses in annealing of ion-implantation damage in Si", Appl. Phys. Lett., 33 (1), 21-3 (1978), EEA81-42341.
455. Smith, B. J., Stephen, J. and Hammersley, P. J., "Doping centres in partially annealed carbon implanted silicon", Radiat. Eff., 26 (1-2), 17-21 (1975), EEA79-8781.
456. Stephen, J., Smith, B. J. and Blamires, N. G., "Pulsed laser annealing of ion implanted silicon", AIP Conf. Proc., no. 50, 337-43 (1978), EEA82-38217.
457. Suski, J., Krynicki, J. and Rzewuski, H., "Ionization enhanced annealing in P and As implanted Si layers", Radiat. Eff., 30 (2), 125-6 (1976), PA79-87920.

458. Takai, M., Tsien, P. H., Tsou, S. C., Roschenthaler, D., Ramin, M., Ryssel, H. and Ruge, I., "CW CO_2-laser annealing of arsenic implanted silicon", Appl. Phys., 22 (2), 129-36 (1980), EEA83-43997.
459. Tamminga, Y., Eggermont, G. E. J., Hofker, W. K., Hoonhout, D., Garrett, R. and Saris, F. W., "Differences between ruby and Nd:YAG laser annealing of ion implanted silicon", Phys. Lett. Vol. A, 69A (6), 436-8 (1979), EEA82-22089.
460. Tatsuta, S. and Sakurai, T., "Low-temperature annealing characteristics of low-energy BF_2^+ implanted silicon", Fujitsu Sci. & Tech. J., 14 (4), 73-82 (1978), EEA82-26747.
461. Tice, W. K., Tan, T. Y. and Geipel, H. J., "The isochronal annealing behavior of silicon implanted with arsenic ions", Electrochemical Society Fall Meeting (Extended abstracts only received), 1976, p. 863-4, EEA80-36253.
462. Tkachev, V. D., Schrodel, C. and Mudryi, A. V., "Annealing of lattice damage in ion implanted silicon", Proceedings of the International Conference on Ion Beam Modification of Materials, 1978, p. 1143-50.
463. Tokuyama, T., Miyao, M. and Yoshihiro, N., "Nature and annealing behavior of disorders in ion implanted silicon", Jpn. J. Appl. Phys., 17 (8), 1301-15 (1978), EEA81-45934.
464. Tseng, W. F. and Mayer, J. W., "Grain size dependence in a self-implanted silicon layer on laser irradiation energy density", Appl. Phys. Lett., 32 (12), 824-6 (1978), EEA81-42217.
465. Tsou, S. C., Tsien, P. N., Takai, M., Roschenthaler, D., Ramin, M., Ryssel, H. and Ruge, I., "Front and back surface CW CO_2-laser annealing of arsenic ion-implanted silicon", Appl. Phys., 23 (2), 163-8 (1980), PA84-2475.
466. Tsu, R., Baglin, J. E., Tan, T. Y., Tsai, M. Y., Park, K. C. and Hodgson, R., "Laser recrystallization of ion-implanted Si by frequency-doubled Nd:YAG laser", AIP Conf. Proc., no. 50, 344-50 (1978), EEA82-38218.
467. Turos, A. and Greek, J., "Pulsed electron-beam annealing of high-dose arsenic implanted silicon", Appl. Phys., 22 (4), 385-8 (1980), EEA83-53644.
468. Uda, K. and Kamoshida, M., "Annealing characteristics of highly p^+-ion implanted silicon crystal-two-step anneal", J. Appl. Phys., 48 (1), 18-21 (1977), EEA80-20910.
469. Van Vechten, J. A., Tsu, R. and Saris, F. W., "Nonthermal pulsed laser annealing of Si, plasma annealing", Phys. Lett. A, 74A (6), 422-6 (1979), EEA83-14322.
470. Venkatesan, T. N., Golovchenko, J. A., Poate, J. M., Cowan, P. and Celler, G. K., "Dose dependence in the laser annealing of arsenic-implanted silicon", Appl. Phys. Lett., 33 (5), 429-31 (1978), EEA81-49713.
471. Wang, K. L., Liu, Y. S., Kirkpatrick, C. G. and Possin, G. E., "Electrical properties and annealing kinetics study of laser-annealed ion-implanted silicon", AIP Conf. Proc., no. 50, 569-78 (1978), EEA82-34284.

472. Welch, J. D., Cobbold, R. S. C. and Davies, J. A., "Migration of ion-implanted krypton in silicon during anneal", J. Appl. Phys., 48 (11), 4540-3 (1977), EEA81-10274.
473. White, C. W., Christie, W. H., Appleton, B. R., Wilson, S. R., Pronko, P. P. and Magee, C. W., "Redistribution of dopants in ion-implanted silicon by pulsed-laser annealing", Appl. Phys. Lett., 33 (7), 662-4 (1978), EEA82-4294.
474. White, C. W., Narayan, J. and Young, R. T., "Laser annealing of ion implanted silicon", AIP Conf. Proc., no. 50, 275-90 (1978), EEA82-34273.
475. White, C. W., Pronko, P. P., Wilson, S. R., Appleton, B. R., Narayan, J. and Young, R. T., "Effects of pulsed ruby-laser annealing on As and Sb implanted silicon", J. Appl. Phys., 50 (5), 3261-73 (1979), EEA82-42454.
476. White, C. W., Wilson, S. R., Appleton, B. R. and Young, F. W., Jr., "Supersaturated substitutional alloys formed by ion implantation and pulsed laser annealing of group-III and group-V dopants in silicon", J. Appl. Phys., 51 (1), 738-49 (1980), EEA83-23913.
477. Williams, J. S., Brown, W. L., Leamy, J. H., Poate, J. M., Rodgers, J. W., Rousseau, D., Rozgonyi, G. A., Shelnutt, J. A. and Sheng, T. T., "Solid-phase epitaxy of implanted silicon by CW Ar ion laser irradiation", Appl. Phys. Lett., 33 (6), 542-4 (1978), EEA82-678.
478. Williams, J. S. and Grant, W. A., "Annealing behaviour of high dose rare-gas implantations in silicon", Applications of Ion Beams to Materials, 1975, p. 31-6, Publ. 1976, EEA79-32920.
479. Wilson, S. R., Appleton, B. R., White, C. W., Narayan, J. and Greenwald, A. C., "Comparison of pulsed electron beam-annealed and pulsed ruby laser-annealed ion-implanted silicon", IEEE Trans. Nucl. Sci., NS 26 (1), pt. 2, 1693-6 (1979), EEA82-26634.
480. Wilson, S. R., White, C. W., Young, F. W., Jr., Appleton, B. R. and Narayan, J., "Nonequilibrium solubility and segregation in ion implanted, laser annealed silicon", J. Phys. Colloq., 41 (C-4), 91-6 (1980), PA83-84296.
481. Wu, C. P. and Schnable, G. L., "Laser annealing to round the edges of silicon structures", RCA Rev., 40 (3), 339-44 (1979), EEA83-9670.
482. Yamazaki, H., Nojima, S. and Harada, H., "Annealing characteristics of heavily-boron implanted layers in silicon", Electr. Commun. Lab. Tech. J., 26 (4), 1261-8 (1977), Japanese, EEA81-10116.
483. Young, R. T., Narayan, J., White, C. W., Wood, R. F. and Cleland, J. W., "Electrical characteristics of ion-implanted laser-annealed silicon", Report CONF-780932-4, Oak Ridge National Lab., TN, 1978, 8 pp.
484. Young, R. T., White, C. W., Clark, G. J., Narayan, J., Christie, W. H., Murakami, M., King, P. W. and Kramer, S. D., "Laser annealing of boron-implanted silicon", Appl. Phys. Lett., 32 (3), 139-41 (1978), EEA81-19416.

c. Channeling

485. Acerbi, E., Birattari, C., Candoni, B., Castiglioni, M. and Succi, C., "Channeling of high-energy protons in Si and W", Nuovo Cimento B, 29B (2), 257-69 (1975), PA79-9325.
486. Baeri, P., Campisano, S. U., Ciavola, G., Foti, G. and Rimini, E., "Radial distribution of ion-induced defects determined by channelling measurements", Nucl. Instrum. & Methods, 132, 237-40 (1976), PA79-53184.
487. Baeri, P., Campisano, S. U., Ciavola, G., Foti, G. and Rimini, E., "Distribution across the channel of defects induced by nitrogen bombardment in silicon", Appl. Phys. Lett., 28 (1), 9-11 (1976), PA79-36457.
488. Bulgakov, Y. V. and Shulga, V. I., "Transparency oscillations of a silicon single crystal in passing from axial to planar channeling", Radiat. Eff., 28 (1-2), 15-21 (1976), PA79-53233.
489. Campisano, S. U. and Ciavola, G., "Channeling and RHEED analyses of Pb-implantation in silicon", Appl. Phys., 15 (2), 233-7 (1978), EEA81-19419.
490. Cembali, F., Galloni, R. and Zignani, F., "Electrical activation processes of P^+ ions channeled along the (110) axis of silicon: effect of annealing on carrier profile shape", Radiat. Eff., 26 (3), 161-71 (1975), PA79-13554.
491. Comas, J. and Wilson, R. G., "Channeling and random equivalent depth distributions of 150 keV Li, Be, and B implanted in silicon", J. Appl. Phys., 51 (7), 3697-701 (1980), PA83-100293.
492. Deutch, B. I., et al, "Pulsed, Q-switched ruby laser annealing of Bi implanted silicon crystals investigated by channeling", Radiat. Eff., 45 (1-2), 103-10 (1979), PA83-59542.
493. Furuya, T., Nishi, H., Inada, T. and Sakurai, T., "Channeled-ion implantation of group-III and group-V ions into silicon", J. Appl. Phys., 49 (7), 3918-21 (1978), EEA81-42348.
494. Gotz, G., Gruska, B. and Hedler, H., "Investigation of extended defects in ion implanted Si and GaAs by the energy dependence of the dechanneling of He^+-ions", Proceedings of the International Conference on Ion Beam Modification of Materials, 1978, p. 1151-60.
495. Grahmann, H., Feuerstein, A. and Kalbitzer, S., "Critical channeling angles of low energy ions in silicon", Radiat. Eff., 29 (2), 117-19 (1976), PA79-68585.
496. Huez, M., Cappellani, F. and Restelli, G., "Channeling effect measurements of arsenic implants in silicon", Radiat. Eff., 32 (3-4), 147-54 (1977), EEA80-28832.
497. Inada, T., Nishi, H. and Furuya, T., "Channeled-ion implantation of B and P ions into Si", Fujitsu Sci. & Tech. J., 13 (3), 69-83 (1977), EEA81-10267.
498. Lecrosnier, D., Paugam, J. and Gallou, J., "Channelling of boron ions into silicon", Appl. Phys. Lett., 30 (7), 323-5 (1977), EEA80-24860.

499. Melvin, J. D. and Tombrello, T. A., "Energy loss of low energy protons channeling in silicon crystals", Radiat. Eff., 26 (1-2), 113-26 (1975), PA79-13548.
500. Seidel, T. E., "Channeling of implanted phosphorus through polycrystalline silicon", Appl. Phys. Lett., 36 (6), 447-50 (1980), EEA83-28915.
501. Turos, A., Wielunski, L., Wielunski, M. and Wojtowicz-Natanson, B., "Channeling analysis of disorder structure in neon implanted silicon", Nucl. Instrum. & Methods, 149 (1-3), 421-4 (1978), EEA82-806.
502. Wada, Y., Nishimatsu, S. and Hashimoto, N., "Arsenic ion channeling through single crystal silicon", J. Electrochem. Soc., 127 (1), 206-10 (1980), EEA83-23915.
503. Williams, J. S., Brown, W. L. and Poate, J. M., "Channeling analysis of CW argon laser annealed, As^+-implanted silicon", AIP Conf. Proc., no. 50, 399-404 (1978), EEA82-38224.

d. Concentration Profiles

504. Abroyan, I. A., Zitta, N. F., Konyshev, V. V., Titov, A. I. and Khlebalkin, A. V., "Depth distribution of structural defects in ion-bombarded silicon", Bull. Acad. Sci. USSR, Phys. Ser., 40 (8), 170-3 (1976), EEA81-1017.
505. Anand, K. V., Sobhy, M. I. and El-Dhaher, A. H. G., "Distribution profiles of boron-implanted layers in silicon using a high resolution anodic oxidation cell", Int. J. Electron., 40 (6), 617-23 (1976), PA79-69518.
506. Bach, K. and Muhle, R., "Determination of doping profiles in ion implanted silicon by means of a scanning electron microscope", Exp. Tech. Phys., 24 (3), 255-65 (1976), German, PA79-80444.
507. Bakowski, A., "A method of determination of the diffusion coefficient from measurements of the depth distribution of carrier concentration in Si and examples of applications to ion implanted diffusion source. Part 1", Proceedings of the International Conference on Ion Beam Modification of Materials, 1978, p. 225-38.
508. Baragiola, R. A., Chivers, D., Doods, D., Grant, W. A. and Williams, J. S., "Ranges in silicon of ions with atomic numbers $62 \leq Z_1 \leq 66$ at 100 keV", Phys. Lett. A, 56A (5), 371-3 (1976), PA79-57723.
509. Baronova, A. S., Zorin, E. I., Karmanov, B. T., Pavlov, P. V. and Tetel'baum, D. I., "Anomalous diffusion of arsenic introduced into silicon by ion implantation", Sov. Microelectron, 7 (1), 37-41 (1978), EEA82-8269.
510. Bayadilov, E. M., Katsaurov, L. N., Krasnopevisev, V. V. and Nurgaliev, K., "Effect of heat treatment on the distribution profile of lithium implanted in silicon at high doses", Sov. Phys. Lebedev Inst. Rep., no. 5, 30-4 (1978), EEA82-42464.

511. Bernhard, F., Kerkow, H. and Kudella, F., "Depth profile measurements of copper in silicon by ion-induced X-ray emission", Proceedings of the International Conference on Ion Beam Modification of Materials, 1978, p. 1733-42.
512. Blank, P., Wittmaack, K. and Schulz, F., "The influence of sputtering, range shortening and stress-induced outdiffusion on the retention of xenon implanted in silicon", Nucl. Instrum. & Methods, 132, 387-92 (1976), PA79-53158.
513. Brylowska, I., Subotowicz, M., Paprocki, K. and Derewiecki, H., "Doping profiles from the capacity-voltage (C-V) characteristics of the [111] Si(p) single crystals doped by the In^+ ion-implantation", Proceedings of the International Conference on Ion Beam Modification of Materials, 1978, p. 287-96, PA83-88203.
514. Burkhardt, F. and Wagner, C., "Depth distribution of phosphorus implanted into silicon", Phys. Status Solidi A, 39 (1), K63-5 (1977), EEA80-20919.
515. Campisano, S. U., Baeri, P., Foti, G., Ciavola, G. and Rimini, E., "Impurity redistribution during laser irradiation in ion implanted silicon", Radiat Eff., 48 (1-4) 187-90 (1980), EEA83-48712.
516. Campisano, S. U., Baeri, P., Grimaldi, M. G., Foti, G. and Rimini, E., "Surface accumulation of Te atoms in laser melted Te-implanted silicon", J. Appl. Phys., 51 (7), 3968-70 (1980), PA83-100258.
517. Christodoulides, C. E., Carter, G. and Williams, J. S., "Implant redistribution in high-dose ion implanted and annealed silicon", Proceedings of the International Conference on Ion Beam Modification of Materials, 1978, p. 305-12, EEA83-53656.
518. Chu, A., "A theoretical approach to the calculation of annealed impurity profiles of ion-implanted boron into silicon", Thesis, Stanford University, 1977, 148 pp., Order no. 77-25651.
519. Chu, W. K., Kastl, R. H. and Murley, P. C., "Low-energy antimony implantation in silicon. I. Profile measurements and calculation", Radiat. Eff., 47 (1-4), 1-6 (1980), EEA83-44003.
520. Cullis, A. G., Webber, H. C. and Chew, N. G., "Crystal growth and dopant redistribution in ion implanted and laser annealed silicon", Electron Microscopy and Analysis Symposium, 1979, p. 145-8, Publ. 1980, EEA83-34815.
521. Dietrich, H. B. and Plew, L. E., "Fluorine-19 range-energy curve in silicon from 100 to 550 keV", Appl. Phys. Lett., 29 (7), 406-8 (1976), CA85-152856.
522. Dietrich, H. B., Weisenberger, W. H. and Comas, J., "Anomalous migration of ion-implanted Al in Si", Appl. Phys. Lett., 28 (4), 182-4 (1976), EEA79-24691.
523. Dikii, N. P., Matyash, P. P. and Skakun, N. A., "Profiles of radiation damage and implanted boron in silicon", Sov. Phys.-Semicond., 9 (3), 395-6 (1975), PA79-17751.
524. Fair, R. B. and Pappas, P. N., "Diffusion of ion-implanted B in high concentration P- and As-doped silicon", J. Electrochem. Soc., 122 (9), 1241-4 (1975), PA79-1825.

525. Fair, R. B. and Tsai, J. C. C., "The diffusion of ion-implanted arsenic in silicon", J. Electrochem. Soc., 122 (12), 1689-96 (1975), PA79-22942.
526. Fair, R. B. and Tsai, J. C. C., "Profile parameters of implanted-diffused arsenic layers in silicon", J. Electrochem. Soc., 123 (4), 583-6 (1976), PA79-58545.
527. Gashtold, V. N., Gerasimenko, N. N., Dvurechenskii, A. V. and Smirnov, L. S., "Profiles of defects produced by the implantation of ions in silicon", Sov. Phys.-Semicond., 9 (5), 551-3 (1975), PA79-14715.
528. Gorelkinskiy, Y. V., Nevinnyi, N. N. and Botvin, V. A., "The spatial distribution of Si interstitial complex produced in silicon by hydrogen ion implantation", Proceedings of the International Conference on Ion Beam Modification of Materials, 1978, p. 1651-60.
529. Gotz, G., Klinge, K. D., Schwabe, F. and Solovjov, V., "Depth profile of radiation damage in ion implanted silicon single crystals", Exp. Tech. Phys., 25 (1), 71-9 (1977), German, EEA80-20907.
530. Grant, W. A., Williams, J. S. and Dodds, D., "Measurements of the lateral spread of heavy ions implanted into silicon", Radiat. Eff., 29 (3), 189-20 (1976), PA79-87917.
531. Harada, H., "Diffusion coefficients and range parameters of ion-implanted boron and phosphorus in silicon", Electrochemical Society Spring Meeting (Extended Abstracts), 1976, p. 349-51 PA79-87926.
532. Hirao, T., Inoue, K. and Takayanagi, S., "Depth distribution of knock-on nitrogen in Si by phosphorus implantation through Si_3N_4 films", Appl. Phys. Lett., 31 (8), 505-8 (1977), EEA81-5914.
533. Hirao, T., Inoue, K., Takayanagi, S. and Yaegashi, Y., "The concentration profiles of projectiles and recoiled nitrogen in Si after ion implantation through Si_3N_4 films", J. Appl. Phys., 50 (1), 193-201 (1979), EEA82-30375.
534. Hirao, T., Inoue, K., Yaegashi, Y. and Takayanagi, S., "The concentration profiles of phosphorus, arsenic and recoiled oxygen atoms in Si by ion implantation into SiO_2-Si", Jpn. J. Appl. Phys., 18 (3), 647-56 (1979), EEA82-26750.
535. Hofker, W. K., Werner, H. W., Oosthoek, D. P. and Koeman, N. J., "Redistribution effects associated with sequential implantations", Applications of Ion Beams to Materials, 1975, p. 13-17, Publ. 1976, PA79-69551.
536. Inoue, K., Hirao, T., Takayanagi, S. and Yaegashi, Y., "Asymmetrical profiles of ion implanted phosphorus in silicon", Jpn. J. Appl. Phys., 18 (2), 367-72 (1979), EEA82-22085.
537. Iwaki, M., Gamo, K., Masuda, K., Namba, S., Ishihara, S., Kimura, I. and Yokota, K., "Experimental method for measuring both atom and carrier concentation profiles in the same sample of ion-implanted silicon layers by radioactive-ion implantation", Nucl. Instrum. & Methods, 127 (1), 93-8 (1975), PA78-72506.

538. Jaskolska, H., Rowinska, L. and Walis, L., "Application of neutron activation analysis for the determination of implantation profiles of phosphorus in semiconductor grade silicon", J. Radioanal. Chem., 38 (1-2), 29-35 (1977), EEA81-5920.
539. Kappert, H. F., Heidemann, K. F., Eichholz, D., te Kaat, E. and Rothemund, W., "Range and range straggling of oxygen implanted into silicon at energies between 2 and 20 MeV", Appl. Phys., 21 (2), 151-8 (1980).
540. Kappert, H. F., Heidemann, K. F., Grabe, B. and te Kaat, E., "Range and damage profiling after heavy ion implantation in the MeV region", Phys. Status Solidi A., 47 (2), 751-62 (1978), PA81-75686.
541. Koyama, H., "Redistribution of implanted oxygen and carbon in silicon", J. Appl. Phys., 51 (6), 3202-5 (1980).
542. Kutukova, O. G., and Strel'tsov, L. N., "Influence of the radiation dose on diffusion processes in ion-implantation-doped layers", Sov. Phys.-Semicond., 10 (4), 475-6 (1976), EEA80-621.
543. Marcyk, G. T. and Streetman, B. G., "Boron impurity profile tailoring in silicon by ion implantation and measurement by glow discharge optical spectroscopy", J. Electrochem. Soc., 123 (9), 1388-91 (1976), PA79-91553.
544. Mende, G., "A comparison of calculated arsenic implantation profiles in silicon with experimental results", Phys. Status Solidi A., 45 (2), K121-3 (1978), EEA81-23874.
545. Morikawa, Y., Takeda, T. and Nagami, K., "Redistribution of P in Si by Ar hot implantation", Jpn. J. Appl. Phys., 16 (7), 1281-2 (1977), EEA80-36230.
546. Motooka, T. and Watanabe, K., "Damage profile determination of ion-implanted Si layers by ellipsometry", J. Appl. Phys., 51 (8), 4125-9 (1980).
547. Mueller, K., Henkelmann, R. and Boroffka, H., "Determination of low dose boron implanted concentration profiles in silicon by the (n,α) reaction", Nucl. Instrum. Methods, 129 (2), 557-9 (1975), CA85-56121.
548. Muller, H., Gyulai, J., Chu, W. K. and Mayer, J. W., "Influence of an oxidizing annealing ambient on the distribution of As, Sb, and Ga implanted into silicon", J. Electrochem. Soc., 122 (9), 1234-8 (1975), PA79-2725.
549. Murarka, S. P., "Diffusion and segregation of ion-implanted boron in silicon in dry oxygen ambients", Phys. Rev. B, 12 (6), 2502-19 (1975), PA79-18900.
550. Mustafin, T. N., Kachurin, G. A., Popov, V. P., Pridachin, N. V., Seryapin, V. G. and Smirnov, L. S., "Diffusion of implanted zinc in laser annealed silicon layers", Proceedings of the International Conference on Ion Beam Modification of Materials, 1978, p. 727-34.
551. Myers, D. R. and Wilson, R. G., "Alignment effects on implantation profiles in silicon", Proceedings of the International Conference on Ion Beam Modification of Materials, 1978, p. 103-12, EEA83-44010.

552. Nagy, A. Z., Bogancs, J., Gyulai, J., Csoke, A., Nazarov, V., Seres, Z., Szabo, A. and Yazvitsky, Yu., "Determination of boron range distribution in ion-implanted silicon by the ^{10}B (n,)^{7}Li reaction", J. Radionanal. Chem., 38 (1-2), 19-27 (1977), EEA81-5919.
553. Nishi, H., Sato, T. and Furuya, T., "Uniform and sloping profiles obtained by multiple dose-energy implantation", Fujitsu Sci. & Tech. J., 11 (4), 113-22 (1975), PA79-36462.
554. Onheiser, P., Zaruba, I., Gervena, J., Hnatowicz, V., Kosina, Z. and Kvitek, J., "A study of boron ion implanted profiles in silicon from ^{10}B(n,)^{7}Li reaction product spectra", Proceedings of the International Conference on Ion Beam Modification of Materials, 1978, p. 449-60.
555. Ota, Y., "N-type doping techniques in silicon molecular beam epitaxy by simultaneous arsenic ion implantation and by antimony evaporation", J. Electrochem. Soc., 126 (10), 1761-5 (1979), EEA83-9698.
556. Perloff, D. S., Wahl, F. E. and Kerr, J. T., "Measurements of the doping uniformity of ion implanted silicon wafers", Electrochemical Society Spring Meeting (Extended Abstracts), 1976, p. 514-16, PA79-91565.
557. Petrov, I. N., Georgiev, V. K. and Glukharov, T. N., "Effect of consecutive high-temperature treatments in inert and oxidative media on the distribution profiles of ion-implanted impurities in silicon", Bulg. J. Phys., 5 (1), 71-5 (1978), EEA81-42344.
558. Petrov, I. N., Tonkov, A. S., Ranev, V. Z. and Grancharov, S. D., "Effect of high-temperature treatments in inert media on the distritution profiles of implanted ^{11}B ions in n-type silicon", Bulg. J. Phys., 4 (1), 58-62 (1977), Russian, EEA81-1193.
559. Pham, M. T., "A very simple method for profiling the ion-implanted Si surface", Phys. Status Solidi A, 49 (1), 261-5 (1978), EEA81-49727.
560. Picraux, S. T., Follstaedt, D. M., Baeri, P., Campisano, S. U., Foti, G. and Rimini, E., "Depth profile studies of extended defects induced by ion implantation in Si and Al", Proceedings of the International Conference on Ion Beam Modification of Materials, 1978, p. 1863-72, PA84-2533.
561. Ricco, R. P., Goldstein, J. I. and McCallum, J. G., "The diffusion of impanted boron in silicon", J. Electrochem. Soc., 124 (2), 276-9 (1977).
562. Ross, R., "Determination of implantation profiles of boron phosphorus, and arsenic in silicon", Zentralinst. Kernforsch., Rossendorf Drsden, ZfK-294, 209-11 (1975), CA86-82540.
563. Rybka, V. and Cerny, F., "Processing parameters for the diffusion redistribution of boron and phosphorus implanted in silicon", TESLA Electron., 11 (2), 56-7 (1978), EEA82-4308.
564. Ryssel, H., Muller, K., Haberger, K., Henkelmann, R. and Jahnel, F., "High concentration effects of ion implanted boron in silicon", Appl. Phys., 22 (1), 35-8 (1980), EEA83-34792.

565. Sadana, D. K., Fletcher, J. and Booker, G. R., "Structural and electrical profiles for double damage layers in ion-implanted silicon", Electron. Lett., 13 (21), 632-3 (1977), EEA81-1019.
566. Sato, T., Nishi, H. and Furuya, T., "Redistributions of implanted boron and phosphorous sources during thermal diffusion", Fujitsu Sci. & Tech. J., 12 (1), 115-24 (1976), EEA79-28396.
567. Schirmer, G., Reiche, H. P., Jahn, U. and Frey, H., "Diffusion behavior of boron implants in silicon", Int. Conf. Ion Implantation Semicond., Proceedings, 1978, p. 195-202, CA90-96400.
568. Schroder, D. K., Braggins, T. T. and Hobgood, H. M., "The doping concentrations of indium-doped silicon measured by Hall, C-V, and junction-breakdown techniques", J. Appl. Phys., 49 (10), 5256-9 (1978), EEA82-8264.
569. Shah, P. and Schwettmann, F. N., "Redistribution of implanted arsenic in silicon", Electrochemical Society Spring Meeting, Extended Abstracts, 1975, p. 437-9, PA79-2764.
570. Shakun, N. A., Dikii, N. P. and Matyash, P. P., "Vertical distribution and annealing of defects in silicon doped with lithium ions", Sov. Phys.,-Solid State, 17 (3), 592-3 (1975), PA79-6413.
571. Sverdlina, D. A., Kuznetsov, Y. N. and Kuzovlev, I. A., "Study of the distribution of doping impurities throughout ion-implanted and epitaxial silicon layers", Prikl. Spektrosk., 198-200 (1977), CA89-139245.
572. Thomas, J. H., "Phosphorus concentration in hydrogenated amorphous silicon using ion-implanted references", J. Vac. Sci. & Technol., 17 (6), 1306-8 (1980).
573. Tsukamoto, K., Akasaka, Y. and Kijima, D., "Thermal diffusion of ion-implanted arsenic in silicon", Jpn. J. Appl. Phys., 19 (1), 87-95 (1980), EEA83-18993.
574. Vaidyanathan, K. V., Chatterjee, P. K. and Streetman, B. G., "Comment on 'Evidence for electronic stopping in ion implantation: shallower profile of lighter isotope ^{10}B in Si'", Appl. Phys. Lett., 27 (12), 648-9 (1975), EEA79-12775.
575. Wada, Y. and Hashimoto, N., "Anomalous carrier profiles in $BF^{+}{}_{2}$- ion-implanted silicon", J. Appl. Phys., 50 (9), 5720-5 (1979), EEA83-743.
576. Wada, Y. and Hashimoto, N., "Anomalous carrier tail generation mechanism in $BF^{+}{}_{2}$ ion-implanted layers of silicon", J. Appl. Phys., 50 (10), 6257-61 (1979), PA83-19819.
577. Wang, J. C., Wood, R. F., White, C. W., Appleton, B. R., Pronko, P. P., Wilson, S. R. and Christie, W. H., "Dopant profile changes induced by laser irradiation of silicon: comparison of theory and experiment", AIP Conf. Proc., no. 50, 123-8 (1978), EEA82-34270.
578. Wilson, R. G. and Comas, J., "Correlation of atomic distribution and implantation induced damage profiles in Be-ion implanted Si", Proceedings of the International Conference on Ion Beam Modification of Materials, 1978, p. 547-56, PA83-88199.

579. Wu, C. P., Douglas, E. C. and Mueller, C. W., "Redistribution of ion-implanted impurities in silicon during diffusion in oxidizing ambients", IEEE Trans. Electron Devices, ED-23 (9), 1095-7 (1976), PA79-87894.
580. Yokota, K., Iwaki, M., Gamo, K., Masuda, K., Namba, S., Ishihara, S. and Kimura, I., "Diffusion of arsenic atoms from ion implanted layers in silicon with various boron concentrations", Applications of Ion Beams to Materials, 1975, p. 6-12, Publ. 1976, PA79-69550.
581. Yokota, K., Tamura, S., Horiuchi, K., Gamo, K., Naniwa, S., Masuda, K., Iwaki, M., Ishihara, S. and Kimura, I., "Diffusion of gallium in arsenic-doped silicon from ion implantation layer", Daigaku Genshiro Jikkensho, KURRI-TR-177, 19-22 (1978), Japanese, CA90-96397.
582. Yusa, A., Yatsurugi, Y. and Takaishi, T., "Rapid determination of the distribution of ^{31}P in neutron-irradiated silicon", J. Electrochem. Soc., 124 (2), 312-14 (1977), EEA80-13456.
583. Ziegler, J. F., Wu, C. P., Williams, P., White, C. W., et al., "Profiling hydrogen in materials using ion beams", Nucl. Instrum. & Methods, 149 (1-3), 19-39 (1978), EEA82-4305.

e. Electrical and Optical Properties

584. Adams, J. R. and Bashara, N. M., "Determination of the complex refractive index profiles in P_{31}^{+} ion implanted silicon by ellipsometry", Surf. Sci., 49 (2), 441-58 (1975), EEA78-28653.
585. Akimchenko, I. P., Kisselva, K. V., Krasnopevtsev, V. V., Touryanski, A. G. and Vavilov, V. S., "Structure and optical properties of silicon implanted by high doses of 70 and 310 keV carbon ions", Radiat. Eff., 48 (1-4), 7-12 (1980), EEA83-53655.
586. Arifov, U. A., Abdullaev, N. and Ayukhanov, A. K., "Change in the anomalously large photovoltage of silicon and germanium films during bombardment by low-energy ions", Izv. Akad. Nauk Uzb. SSR. Ser. Fiz.-Mat. Nauk, no. 2, 38-41 (1976), Russian, CA85-12779.
587. Beanland, D. G. and Chivers, D. J., "Sheet resistance variation on color-banded silicon following high dose implantations at high dose rates", J. Vac. Sci. & Technol., 15 (4), 1536-40 (1978), EEA81-49723.
588. Beanland, D. G. and Chivers, D. J., "Color-band generation during the high dose ion implantation of silicon wafers", J. Electrochem. Soc., 125 (8), 1331-8 (1978), EEA82-803.
589. Beanland, D. G., Temple, W. and Chivers, D. J., "Sheet resistance variations of phosphorus implanted silicon at elevated temperatures", Solid-State Electron, 21, 357-60 (1978).
590. Bhattacharya, R. S., Van Der Veen, J. F., Kerkdijk, C. B. W. and Saris, F. W., "Bombardment induced light emission from silicon, silicon nitride and silicon carbide surfaces", Radiat. Eff., 32 (1-2), 25-31 (1977), EEA80-24878.

591. Blank, P. and Wittmaack, K., "The influence of implanted xenon on the sputtering yield of silicon", Radiat. Eff., 27 (1-2), 29-33 (1975), CA84-143491.

592. Bocanek, L., "Sheet resistivity and Hall-effect measurement of implanted silicon by DC and AC methods", Scr. Fac. Sci. Nat. Univ. Parkynianae Brunensis Phys., 4 (1), 39-51 (1974), EEA79-8796.

593. Bourgoin, J. C., Frossati, G., Ravex, A., Thoulouze, D., Vandorpe, M. and Waksmann, B., "Conduction at low temperature in ion-implanted silicon", Phys. Status Solidi B, 92 (2), 585-94 (1979), EEA82-30246.

594. Cullis, A. G., Webber, H. C. and Chew, N. G., "Correlation of the structure and electrical properties of ion-implanted and laser-annealed silicon", Appl. Phys. Lett., 36 (7), 547-50 (1980), PA83-70855.

595. Dekhtyar, I. Y. and Sakharova, S. G., "Positron annihilation in ion-implantation-doped silicon single crystals", Sov. Phys. -Solid State, 20 (4), 625-8 (1978), PA82-15555.

596. Dexter, R. J., "Effects of high dose nitrogen implantation in silicon", Electrochemical Society Spring Meeting (Extended Abstracts), 1976, p. 511-13, PA79-91564.

597. Dietrich, H. B., "Anomalous migration of ion-implanted Al, Ga and In in Si", Electrochemical Society Spring Meeting (Extended Abstracts), 1976, p. 356-7, PA79-87929.

598. Frey, H., Jahn, U., Mittenbacher, J. and Schirmer, G., "Post-implantation of boron-diffused silicon layers with boron-11 ions", Int. Conf. Ion Implantation Semicond., Proc., 1978, p. 203-9, CA90-96401.

599. Frey, H. and Mittenbacher, J., "Influence of electrically active lattice defects on the temperature coefficient of resistance of ion implanted silicon layers", Exp. Tech. Phys., 24 (3), 267-76 (1976), German, EEA79-41320.

600. Galloni, R., Pedulli, L. and Zignani, F., "Study of electrical activity recovery stages in phosphorus implanted silicon", Radiat. Eff., 32 (3-4), 223-7 (1977), EEA80-28834.

601. Garcia, M., Gailliard, J. P. and Roussin, J. C., "Implantation of high doses of antimony for application to the fine epitaxy of silicon on buried layers", Vide, 31 (185), 182-4 (1976), French, EEA80-41402.

602. Gat, A., Gibbons, J. F., Magee, T. J., Peng, J., Deline, V. R., Williams, P. and Evans, C. A., Jr., "Physical and electrical properties of laser-annealed ion-implanted silicon", Appl. Phys. Lett., 32 (5), 276-8 (1978), EEA81-28257.

603. Golja, B. and Nassibian, A. G., "Ar ion implant damage gettering of generation impurities in silicon employing voltage ramping and nitrogen backscattering", IEE J. Solid-State and Electron Devices, 3 (5), 127-32 (1979), EEA82-47025.

604. Grahmann, H. and Kalbitzer, S., "Nuclear and electronic stopping powers of low energy ions with $Z \leq 10$ in silicon", Nucl. Instrum. & Methods, 132, 119-23 (1976), PA79-53210.

605. Grob, A., Grob, J. J. and Siffert, P., "Energy loss and straggling of heavy ions by nuclear interactions in silicon", Nucl. Instrum. & Methods, 132, 273-9 (1976), PA79-53223.
606. Hara, K., "Effects of low dose ion implantation on minority carrier lifetime in dislocation-free p-type silicon", Oyo Buturi, 49 (1), 17-22 (1980), Japanese, EEA83-23625.
607. Hart, R. R., Anderson, C. L., Dunlap, H. L., Seliger, R. L. and Wang, V., "High current density Ga^+ implantations into Si", Appl. Phys. Lett., 35 (11), 865-7 (1979), EEA83-14278.
608. Hlavka, J., "Excess carrier recombination in implanted silicon", Proceedings of the International Conference on Ion Beam Modification of Materials, 1978, p. 357-62.
609. Hubler, G. K., Malmberg, P. R., Carosella, C. A., et al, "Optical effects resulting from deep implants of silicon with N and P", Proceedings of the International Conference on Ion Beam Modification of Materials, 1978, p. 1321-32.
610. Hubler, G. K., Waddell, C. N., Spitzer, W. G., Fredrickson, J. E., Prussin, S. and Wilson, R. G., "High-fluence implantations of silicon: layer thickness and refractive indices", J. Appl. Phys., 50 (5), 3295-3303 (1979), EEA82-42455.
611. Huller, J. and Pham, M. T., "Characterization of ion implanted silicon by the electrolytic reverse current", Phys. Status Solidi A, 43 (2), 505-12 (1977), EEA81-5732.
612. Igras, E., "Postbombardment effects in silicon donor and acceptor implanted layers", Proceedings of the International Conference on Ion Beam Modification of Materials, 1978, p. 391-400.
613. Igras, E. and Rybinski J., "Piezoresistive properties of boron implanted silicon layers", Proceedings of the International Conference on Ion Beam Modification of Materials, 1978, p. 941-8.
614. Karmanov, V. T., Pavlov, P. V., Zorin, E. I., Khokhlov, A. F. and Tetelbaum, D. I., "Absence of donor properties of phosphorus diffusing in ion-bombarded silicon", Sov. Phys.-Semicond., 9 (9), 1171-2 (1975), PA79-54474.
615. Karmanov, V. T., Pavlov, P. V., Zorin, E. I. and Tetel'baum, D. I., "Effect of argon ion bombardment on the diffusion and electrical activity of phosphorus in silicon", Mikroelektronika, 6 (1), 84-7 (1977), CA86-132122.
616. Kassabov, J. D., Voutov, M. P. and Velchev, N. B., "Some characteristics and physical properties of boron implanted high-resistivity layers in silicon", C. R. Acad. Bulg. Sci., 28 (9), 1167-70 (1975), PA79-36460.
617. Kassabov, J. D., Voutov, M. P., Velchev, N. B. and Petrov, I. N., "Ion-implanted p-type high-resistivity layers with high temperature treatment", C. R. Acad. Bulg. Sci., 29 (4), 481-3 (1976), EEA79-37412.
618. Khaibullin, I. B., Shtyrkov, E. I., Zaripov, M. M., Bayazitov, R. M. and Aganov, R. V., "Influence of laser annealing regimes on electrical characteristics of implanted silicon layers", Proceedings of the International Conference on Ion Beam Modification of Materials, 1978, p. 735-50.

619. Kirkpatrick, C. G., Noonan, J. R. and Streetman, B. G., "Recombination luminescence from ion implanted silicon", Radiat. Eff., 30 (2), 97-106 (1976).
620. Korol, V. M. and Zastavnyi, A. V., "Behavior of potassium implanted in silicon", Sov. Phys.-Semicond., 11 (8), 926-9 (1977), EEA81-19277.
621. Kucirkova, A., "Optical reflectance study of damage produced by nitrogen and tellurium ions implanted into silicon", Scr. Fac. Sci. Nat. Univ. Purkynianae Brunensis Phys., 4 (2), 65-78 (1974), EEA79-8799.
622. Ludman, J., Roosild, S. and Vickers, V., "TSC measurements of coupled levels in ion implanted Si", J. Electron. Mater., 7 (4), 535-46 (1978), EEA81-45821.
623. Mahin, A. I., Tetel'baum, D. I. and Khokhlov, A. F., "Electrical conductivity of silicon made amorphous by bombardment with very large ion doses", Sov. Phys.-Semicond., 12 (3), 253-5 (1978), EEA82-8126.
624. Michel, H., "Measurement of the differential capacitance in ion-implanted silicon electrodes", Zentralinst. Kernforsch., Rossendorf Dresden, ZfK-294, 199-200 (1975), German, CA86-82538.
625. Mittenbacher, J. and Frey, H., "Effect of electrically active lattice defects on the temperature coefficient of resistance of ion-implanted silicon layers", Exp. Tech. Phys., 24 (3), 267-76 (1976), German, CA85-102873.
626. Miyao, M., Miyazaki, T. and Tokuyama, T., "Optical reflectivity studies of damage in ion implanted silicon", Jpn. J. Appl. Phys., 17 (5), 955-6 (1978), EEA81-33271.
627. Miyao, M., Yoshihiro, N., Tokuyama, T. and Mitsuishi, T., "Corelation between lattice damage and electrical activation of phosphorus-implanted silicon", J. Appl. Phys., 49 (4), 2573-5 (1978), EEA81-38119.
628. Miyao, M., Yoshihiro, N., Tokuyama, T. and Mitsuishi, T., "Damage-dependent electrical activation of ion-implanted silicon. I. Experiments on phosphorus implants", J. Appl. Phys., 50 (1), 223-30 (1979), EEA82-30239.
629. Motamedi, M. E., Das, P. and Bharat, R., "Surface study of ion-implanted silicon wafers using SAW convolver", J. Appl. Phys., 48 (6), 2135-8 (1977), EEA80-36223.
630. Muller, G. and Kalbitzer, S., "Electrical conductivity of ion implanted amorphous silicon", 4th International Conference on the Physics of Non-Crystalline Solids, 1976, p. 278-83, EEA81-14521.
631. Murakami, K., Masuda, K., Gamo, K. and Namba, S., "Effects ion-implanted atoms upon conduction electron spin resonance (CESR) in a Si:P system", Appl. Phys. Lett., 30 (6), 300-2 (1977), PA80-36133.
632. Murakami, K., Namba, S., Gamo, K. and Masuda, K., "Zeeman energy transfer at the surface layer of P^+ ion implanted silicon", Surf. Sci., 86 (2), 655-64 (1979), PA83-11247.

633. Murakami, K., Namba, S., Kishimoto, N., Masuda, K. and Gamo, K., "Resistance changes induced by electron-spin resonance in ion-implanted Si:P system", J. Appl. Phys., 49 (4), 2401-6 (1978), PA81-68377.
634. Myers, D. R. and Phillips, W. E., "Observation of unequal densities for sulfur defects in silicon predeposited by low fluence ion implantation", J. Electron. Mater., 8 (6), 781-8 (1979), EEA83-9579.
635. Myers, D. R., Roitman, P. and Mayo, S., "Electrical characterization of low-dose ion-implanted silicon annealed with microsecond laser pulses", AIP Conf. Proc., no. 50, 563-8 (1978), EEA82-34283.
636. Myers, D. R. and Streetman, B. G., "Detection by low temperature photoluminescence of oxygen recoils in 'through-oxide' arsenic implanted silicon", Solid State Commun., 18 (7), 815-17 (1976), EEA79-24740.
637. Nakashima, H., Shiraki, Y. and Miyao, M., "Photoluminescence study of laser annealing in phosphorus-implanted and unimplanted silicon", J. Appl. Phys., 50 (9), 5966-9 (1979), EEA83-617.
638. Nishi, H., Sakurai, T. and Furuya, T., "Electrical activation of implanted arsenic in silicon during low temperature anneal", J. Electrochem. Soc., 125 (3), 461-6 (1978), EEA81-28264.
639. Nojima, S., "Anomaly of electrical activation of As atoms implanted in Si by additional Ne irradiation", Jpn. J. Appl. Phys., 17 (8), 1453-4 (1978), EEA81-45820.
640. Panteleev, V. A., Chernyakhovskii, V. V., Ershov, S. N. and Volkov, E. F., "Effect of ion bombardment on the silicon work function", Sov. Phys.-Solid State, 17 (5), 1006-7 (1975), PA79-22414.
641. Petrov, I. N., Georgiev, V. K. and Nencheva, G. P., "Effect of high-temperature treatment in inert atmosphere upon the sheet resistance of ^{11}B ion-implanted layers in n-type silicon", Bulg. J. Phys., 3 (1), 80-5 (1976), EEA79-45948.
642. Pham, M. T. and Hueller, J., "Ion implanted silicon-electrolyte interface", J. Appl. Electrochem., 7 (6), 531-7 (1977), EEA81-45904.
643. Revesz, P., Farkas, G. and Gyulai, J., "Behavior of antimony above solid solubility in silicon produced by implantation and laser annealing", Proceedings of the International Conference on Ion Beam Modification of Materials, 1978, p. 871-86.
644. Ricco, R. P., Goldstein, J. I. and McCallum, J. G., "The diffusion of implanted boron in silicon", J. Electrochem. Soc., 124 (2), 276-9 (1977), EEA80-13455.
645. Romanov, O. V., Uritskii, V. Y. and Yafyasov, A. M., "Influence of ion implantation on surface transport of carriers in silicon", Sov. Phys.-Semicond., 10 (2), 198-201 (1976), EEA79-41473.

646. Romanov, O. V. and Yafyasov, A. M., "Effect of the potential barrier shape on hole transport in a surface inversion channel [low-dose ion implantation]", Sov. Microelectron., 8 (3), 198-202 (1979), EEA83-29276.

647. Ryssel, H., Kranz, H., Bayerl, P. and Schmiedt, B., "Minority carrier lifetime and backscattering measurements of ion-gettered silicon", Proceedings of the International Conference on Ion Beam Modification of Materials, 1978, p. 461-78.

648. Ryssel, H., Kranz, H., Muller, K., Henkelmann, R. A. and Biersack, J., "Comparison of range and range straggling of implanted ^{10}B and ^{11}B in silicon", Appl. Phys. Lett., 30 (8), 399-401 (1977), EEA80-28819.

649. Sanders, I. R., Williams, B. D., Smith, B. J., Stephen, J. and Hinder, G. W., "The effect of argon implantation on the conductivity of boron implanted silicon", Solid-State Electron., 20 (8), 703-7 (1977), EEA80-41398.

650. Scarpone, L. M. and Chlipala, J. D., "Theoretical estimates of the sheet resistance of gaussian n-type ion-implanted layers in semiconductors: phosphorus in silicon", Solid State Electron, 22 (6), 559-66 (1979).

651. Selim, F. A., Blais, P. D., Rai-Choudbury, P. and Yut, R. F., "Electrical characterization of neutron transmutation doped silicon", Electrochemical Society Spring Meeting, Extended Abstracts, 1977, p. 440-2, EEA81-5936.

652. Spangenberg, B., Mladenov, G. M., Orlinov, V. I. and Kaparaschev, S. T., "Dependence of the layer resistance on the regime of boron implantation in silicon and on the annealing conditions", C. R. Acad. Bulg. Sci., 33 (3), 325-7 (1980), German, EEA83-48695.

653. Szabo, A. and Bogancs, J., "Correlation between the energy of implantation and quantity of boron at the determination of boron in silicon", Radiochem. & Radioanal. Lett., 31 (6), 389-95 (1977), EEA81-19428.

654. Tajima, M., "Characterization of neutron-transmutation doping in silicon by the photoluminescence technique", Appl. Phys. Lett., 35 (3), 242-3 (1979), EEA83-4242.

655. Tamura, M., Yagi, K., Sakudo, N., Tokiguchi, K. and Tokuyama, T., "Effects of high dose rate phosphorus implantation into silicon", Jpn. J. Appl. Phys., 17 (suppl. 17-1), 193-200 (1977), EEA81-45933.

656. Thompson, D. A., Chan, M. C. and Campbell, A. B., "Nitrogen implanted silicon carbide: correlation of channeling and electrical studies", Can. J. Phys., 54 (6), 626-32 (1976), CA84-188391.

657. Truche, R., Fourrier, J. Y., Dumontet, M. and Pestie, J., "Some physical and electrical problems of ion implantation at low energy into silicon", Proceedings of the International Conference on Ion Beam Modification of Materials, 1978, p. 537-46.

658. Tsai, M. Y., "Studies of high-dose ion implantation in silicon", Report AD-A069779/7SL, Illinois Univ. at Urbana, Urbana, IL, 1978, 145 pp.
659. Tsai, M. Y., Streetman, B. G., Williams, P. and Evans, C. A., Jr., "Anomalous migration of flourine and electrical activation of boron in BF_2^+-implanted silicon", Appl. Phys. Lett., 32 (3), 144-7 (1978), EEA81-19256.
660. Tsuchimoto, T., Itoh, K. and Kumagaya, F., "Thermal history of sheet resistivity in arsenic implanted silicon", Solid-State Electron., 19 (10), 892-3 (1976), EEA79-45955.
661. Varahramyan, K., Webster, R. T., Das, P. and Bharat, R., "Contactless monitoring of impurity activation in ion-implanted silicon by surface acoustic wave techniques", J. Appl. Phys., 51 (2), 1234-7 (1980), EEA83-29136.
662. Vasilev, V., Amov, B. and Dzhakov, A., "Surface potential of silicon implanted with boron and phosphorus", Int. Conf. Implantation Semicond., Proc., 1978, p. 476-85, CA90-131608.
663. Wang, K. L., "Defect spacial distributions in annealed ion-implanted silicon measured by a transient capacitance technique", Appl. Phys. Lett., 29 (11), 200-2 (1976).
664. Wang, K. L., Liu, Y. S. and Burman, C., "Relationships of electrical properties and melting threshold in laser-annealed ion-implanted silicon", Appl. Phys. Lett., 35 (3), 263-5 (1979), EEA83-4243.
665. Wang, K. L., Liu, Y. S., Kirkpatrick, C. G. and Possin, G. E., "Electrical properties and annealing kinetics study of laser-annealed ion-implanted silicon", AIP Conf. Proc., no. 50, 569-78 (1978), EEA82-34284.
666. Wesch, W. and Gotz, G., "Influence of ion implantation on the optical properties of silicon", Proceedings of the International Conference on Ion Beam Modifications of Materials, 1978, p. 1367-77.
667. Wiggers, L. W. and Saris, F. W., "Effect of the Fermi level position in silicon on ion-induced displacement of impurities", Proceedings of the International Conference on Ion Beam Modification of Materials, 1978, p. 583-98.
668. Yagi, K., Oyo, K., Tamura, M. and Tokuyama, T., "Carrier-concentration reduction in high-dose phosphorus-implanted silicon caused by wet-oxygen oxidation", Appl. Phys. Lett., 35 (3), 257-7 (1979), EEA83-4244.
669. Young, R. T., Narayan, J., White, C. W., Wood, R. F., Cleland, J. W., Westbrook, R. D. and Mooney, P. M., "Electrical characteristics of ion-implanted laser-annealed silicon", Radiat. Eff., 47 (1-4), 4-8 (1980), EEA83-43887.
670. Zastavnyi, A. V., Korol, V. M., Mikhaleva, A. N. and Prozorovskii, V. D., "Temperature dependence of the Hall effect in silicon doped by implantation of sodium ions", Sov. Phys.-Semicond., 13 (5), 580-2 (1979), EEA83-4268.

f. Lattice Location

671. Battaglini, G., Della Mea, G. and Foti, G., "Lattice location of light implanted ions in Si and Ge after laser annealing", Proceedings of the International Conference on Ion Beam Modification of Materials, 1978, p. 669-74.
672. Beloshitsky, V. V., Dikii, N. P., Kumakhov, M. A., Matyash, P. P. and Skakun, N. A., "Use of the channeling technique and the theory of flux peaking effect to determine the location of B in Si", Radiat. Eff., 25 (3), 167-73 (1975).
673. Christodoulides, C. E., Grant, W. A. and Williams, J. S., "Lattice location of lead implanted into silicon at room temperature", Appl. Phys., 13 (4), 391-3 (1977), EEA80-32379.
674. Chu, W. K. and Masters, B. J., "Channeling study of the formation of arsenic clusters in silicon", AIP Conf. Proc., no. 50, 305-10 (1979), EEA82-34275.
675. Dikii, N. P., Matyash, P. P., Svetashev, P. A., Skakun, N. A., Vasilev, V. K., Zorin, E. I., Pavlov, P. V. and Tetelbaum, D. U., "Lattice location of phosphorus atoms implanted into silicon", Phys. Status Solidi A, 32 (2), K165-7 (1975), EEA79-12844
676. Foti, G., Campisano, S. U., Rimini, E. and Vitali, G., "Lattice location of Te in laser-annealed Te-implanted silicon", J. Appl. Phys., 49 (4), 2569-71 (1978), EEA81-38118.
677. Foti, G. and Della Mea, G., "Lattice location of boron implanted silicon after laser annealing", Lett. Nuovo Cimento, 21 (ser. 2, no. 3), 89-93 (1978), EEA81-19427.
678. Gotz, G. and Sommer, G., "Lattice location and ionisation induced annealing of self-interstitials in boron-implanted silicon by Rutherford backscattering", Phys. Status Solidi A, 32 (2), K151-4 (1975), PA79-26034.
679. Hasegawa, S., Ishiwara, H., Furukawa, S. and Shimiza, T., "The lattice location of phosphorus atoms implanted into silicon", Jap. J. Appl. Phys., 15 (2), 391-2 (1976), EEA79-16635.
680. Kolai, E., Lohner, T., Manuaba, A., Mezey, G., Coussement, R., Dezsi, I. and Langouche, G., "Lattice location of Co implanted in silicon", Radiat. Eff., 47 (1-4), 153-5 (1980), PA83-88201.
681. Kryuchkov, Y. Y., Timoshnikov, Y. A., Chernov, I. P., Slavin, N. V. and Azikov, B. S., "Determination of the location of oxygen implanted in the silicon lattice", Sov. Phys.-Semicond., 11 (7), 829-30 (1977), EEA81-14660.
682. Nagy, T., Mezey, G., Kotai, E., Gyulai, J. and Revesz, P., "Lattice location and electrical behavior of group III and V elements implanted into silicon", Report KFKI-76-24, Hungarian Acad. Sci., Budapest, 1976, 6 pp., EEA79-41425.
683. White, C. W., Pronko, P. P., Appleton, B. R., Wilson, S. R. and Narayan, J., "Lattice location of implanted As and Sb in silicon after pulsed laser annealing", Radiat. Eff., 47 (1-4), 31-5 (1980), EEA83-44005.

684. Wilson, S. R., White, C. W., Pronko, P. P., Young, R. T. and Appleton, B. R., "Lattice location of As and Sb implanted in silicon after annealing with a pulsed ruby laser", AIP Conf. Proc., no. 50, 351-6 (1978), EEA82-38219.

g. Modeling

685. Beanland, D. G., "A new model to explain the colours generated on the surface of ion implanted silicon wafers", Radiat. Eff., 33 (4), 219-20 (1977), EEA80-41396.
686. Dennis, J. R. and Hale, E. B., "Crystalline to amorphous transformation in ion-implanted silicon: a composite model", J. Appl. Phys., 49 (3), pt. 1, 1119-27 (1978), EEA81-33267.
687. Desalvo, A., Rosa, R. and Zigma, F., "Dose dependence of penetration and damage profiles of p^+-annealed ions in silicon simulated by computer", Radiat. Eff., 27 (1-2), 89-95 (1975), PA79-35492.
688. Fair, R. B., "Modeling laser-induced diffusion of implanted arsenic in silicon", J. Appl. Phys., 50 (10), 6552-5 (1979), EEA83-9697.
689. Liau, Z. L., Tsaur, B. Y., Lau, S. S., Golecki, I. and Mayer, J. W., "Redistribution of very thin Sb-layers in Si after laser annealing: a theoretical model and preliminary experimental results", AIP Conf. Proc., no. 50, 105-10 (1978), EEA82-34268.
690. Matsumura, H. and Furukawa, S., "Simple formulation for energy straggling of helium in silicon", Radiat. Eff., 27 (3-4), 245-6 (1976), CA85-39830.
691. Merli, G. G., "Computer simulation of electron beam annealing of silicon", Optik, 56 (3), 207-24 (1980), EEA83-48574.
692. Morozov, N. P., Tetelbaum, D. I., Pavlov, P. V. and Zorin, E. I., "Diffusion-coagulation model of accumulation of radiation defects during ion bombardment of silicon", Sov. Phys.-Semicond., 9 (12), 1483-5 (1975), PA79-68543.
693. Morozov, N. P., Tetelbaum, D. I., Pavlov, P. V. and Zorin, E. I., "The calculation of secondary defect formation at ion implantation of silicon", Phys. Status Solidi A, 37 (1), 57-64 (1976), EEA79-46059.
694. Nelson, R. S., "A model for the build-up of disordered material in ion bombarded Si", Radiat. Eff., 32 (1-2), 19-24 (1977), EEA80-24707.
695. Scarfone, L. M. and Chlipala, J. D., "Theoretical estimates of the sheet resistance of gaussian n-type ion-implanted layers in semiconductors: phosphorus in silicon", Solid-State Electron., 22 (6), 559-66 (1979), EEA82-42348.
696. Schultz, J. C. and Collins, R. J., "A computer simulation of laser annealing silicon at 1.06 μm", Appl. Phys. Lett., 34 (1), 84-7 (1979), EEA82-22094.

697. Wang, J. C., Wood, R. F., and Pronko, P. D., "Theoretical analysis of thermal and mass transport in ion-implanted laser-annealed silicon", Appl. Phys. Lett., 33 (5), 455-8 (1978), CA89-172509.
698. Wu, C. P., Douglas, E. C. and Mueller, C. W., "Redistribution of ion-implanted impurities in silicon during diffusion in oxidizing ambients", IEEE Trans. Electron Devices, ED-23 (9), 1095-7 (1976), CA85-185601.

h. Polysilicon

699. Andrews, J. M., "Electrical conduction in implanted polycrystalline silicon", J. Electron. Mater., 8 (3), 227-47 (1979), EEA82-38049.
700. Dvurechenskii, A. V., Kalinin, V. V. and Potapova, L. P.. "Efficiency of defect build-up in ion-implanted polycrystalline silicon films", Phys. Status Solidi A, 54 (2), 457-62 (1979), EEA83-2054.
701. Fujiwara, K., Ohtani, M., Kanayama, K. and Ogata, H., "Quantitative Auger analysis of ion-implanted boron and arsenic in polycrystalline silicon", Surf. Sci., 61 (2), 435-42 (1976), EEA80-20712.
702. Kisielewicz, M., Zielinska-Szot, M. and Zuk, W., "Ion implantation of impurities into polycrystalline silicon", Acta Phys. Pol. A, A56 (5), 609-18 (1979), EEA83-9690.
703. Roulet, M. E., Dutoit, M., Luthy, W. and Affolter, K., "Electrical properties of CW laser-annealed ion-implanted polycrystalline silicon", Appl. Phys. Lett., 37 (8), 737-9 (1980).
704. Seto, J. Y. W., "Annealing characteristics of boron- and phosphorus-implanted polycrystalline silicon", J. Appl. Phys., 47 (12), 5167-70 (1976), EEA80-13452.
705. Shimizu, R., Kang, S. T., Koshikawa, T., Ogata, H., Kanayama, K., Ogata, Y., Akasaka, Y. and Horie, K., "Monte Carlo simulation of depth and lateral profiles of boron atoms implanted in polycrystalline silicon", J. Appl. Phys., 48 (4), 1745-7 (1977), EEA80-24871.
706. Sielanko, J., Krylow, J., Sowa, M. and Zuk, W., "Diffusion of phosphorus implanted in polycrystalline silicon", Int. Conf. Ion Implantation Semicond., Proc., 1978, p. 186-94, CA90-96399.
707. Tsukamoto, K., Akasaka, Y. and Horie, K., "Arsenic implantation into polycrystalline silicon and diffusion to silicon substrate", J. Appl. Phys., 48 (5), 1815-21 (1977), CA86-198803.
708. VanVechten, J. A., "Means to ion implant thin polycrystalline silicon gates without implanting ions into insulating layer beneath gate", IBM Tech. Disclosure Bull., 20 (6), 2454 (1977), EEA82-4297.

709. Wada, Y. and Nishimatsu, S., "Grain growth mechanism of heavily phosphorus-implanted polycrystalline silicon", J. Electrochem. Soc., 125 (9), 1499-504 (1978), EEA82-12567.
710. Wu, C. P. and Magee, C. W., "Laser annealing of ion-implanted polysilicon films", AIP Conf. Proc., no. 50, 357-64 (1978), EEA82-38220.
711. Wu, C. P. and Magee, C. W., "Pulsed laser annealing of ion-implanted polycrystalline silicon films", Appl. Phys. Lett., 34 (11), 737-9 (1979), EEA82-38214.
712. Yaron, G., "Characterization of phosphorus implanted low pressure chemical vapor deposited polycrystalline silicon", Solid-State Electron., 22 (12), 1017-23 (1979), EEA83-28927.

i. Radiation Damage

713. Beanland, D. G. and Chivers, D. J., "Non-uniform damage generation in high dose implantation of silicon at high dose rates", Proceedings of the International Conference on Ion Beam Modification of Materials, 1978, p. 267-76.
714. Cembali, F., Dori, L., Galloni, R., Servidori, M. and Zignani, F., "Radiation damage in silicon produced by phosphorus implantation: random and aligned implants", Radiat. Eff., 36 (1-2), 111-17 (1978), EEA81-28422.
715. Chu, W. K., Kastl, R. H., Lever, R. F., Mader, S. and Masters, B. J., "Radiation damage of 50-250 keV hydrogen ions in silicon", In: Ion Implantation in Semiconductors, 1976, Chernow, F., Borders, J. A. and Brice, D. K., (Eds.), Plenum Publishing Co., New York, p. 483-91 (1977).
716. Chu, W. K., Kastl, R. H., Lever, R. F., Mader, S. and Masters, B. J., "Distribution of irradiation damage in silicon bombarded with hydrogen", Phys. Rev. B, 16 (9), 3851-9 (1977), EEA81-14658.
717. Csepregi, L., Chu, W. K., Muller, H., Mayer, J. W. and Sigmon, T. W., "Influence of thermal history on the residual disorder in implanted (111) silicon", Radiat. Eff., 28 (3-4), 227-33 (1976), PA79-62375.
718. Csepregi, L., Kennedy, E. F., Lau, S. S., Mayer, J. W. and Sigmon, T. W., "Disorder produced by high-dose implantation in Si", Appl. Phys. Lett., 29 (10), 645-8 (1976), EEA80-6574.
719. Das, G., "Analysis of defects in P-implanted silicon", Electrochemical Society Spring Meeting, Extended Abstracts, 1977, p. 190, EEA81-5927.
720. Druzhkov, A. P., Khaibullin, I. B. and Bayazitov, R. M., "Annihilation of positrons in ion-implantation-doped silicon", Sov. Phys.-Semicond., 11 (4), 449-50 (1977), PA81-14380.
721. Dvurechensky, R. V., Rjazantsev, I. A. and Smirnov, L. S., "Defects and impurities interaction enhanced by ionization in ion-implanted silicon", Radiat. Eff., 49 (1-3), 13-15 (1980), PA84-2501.

722. Etlinger, B., Urli, N. B. and Spoglia, U., "Study of defects in p-type silicon implanted with channeled low energy phosphorous ions", Fizika, 9 (3), 143-52 (1977), EEA80-41217.
723. Gavrilenko, V. I., and Litovchenko, V. G., "An investigation of the defect formation processes in ion-implanted silicon", Proceedings of the International Conference on Ion Beam Modification of Materials, 1978, p. 1057-66.
724. Geipel, H. J., and Tice, W. K., "Critical microstructure for ion-implantation gettering effects in silicon", Appl. Phys. Lett., 30 (7), 325-7 (1977), EEA80-24861.
725. Gerasimov, A. B., Golubkov, V. B., Kuteliya, E. P., Mineev, V. P., Mkrtychyan, E. M. and Tsertsvadze, A. A., "Effect of ultraviolet irradiation during the thermal annealing of silicon with implanted boron", Sov. Tech. Phys. Lett., 6 (1), 25-6 (1980).
726. Glaser, E., Gotz, G., Wesch, W. and Frey, H., "Production of radiation defects in silicon at different temperatures", Proceedings of the International Conference on Ion Beam Modification of Materials, 1978, p. 1067-78.
727. Goodwin, C. A., "Prevention of dislocation arrays in thermally annealed, high dose, phosphorus-implanted Si by prior laser annealing", AIP Conf. Proc., no. 50, 411-18 (1978), EEA82-38226.
728. Gotz, G., Hehl, K., Schwabe, F. and Glaser, E., "Investigation of radiation damage in silicon by a backscattering method", Radiat. Eff., 25 (1), 27-32 (1975), PA79-5473.
729. Gotz, G. and Glaser, E., "Radiation damages in argon-implanted and phosphor-implanted silicon single crystals", Wiss, Z. Friedrich-Schiller Univ. Jena, Math. Naturwiss. Reihe, 25 (4), 491-510 (1976), German, EEA80-13461.
730. Gotz, G., Klinge, K.-D., Schwabe, F. and Solovjov, V., "Depth profile of radiation damage in ion implanted silicon single crystals", Exp. Tech.. Phys., 25 (1), 71-9 (1977), German, EEA80-20907.
731. Grotzschel, R., Kalbes, R., Kreissig, U., Rudinger, J. and Schmidt, A., "Recoil implantation and its influence on the damage distribution in silicon", Proceedings of the International Conference on Ion Beam Modification of Materials, 1978, p. 65-74.
732. Hofker, W. K., Josquin, W. J. M. J., Oosthoek, D. P. and Gijsbers, J. R. M., "The influence of the implantation temperature on the generation of dislocations in antimony-implanted and annealed silicon", Proceedings of the International Conference on Ion Beam Modification of Materials, 1978, p. 1121-30
733. Hokari, Y., "Secondary defect generation suppression and diffusivity restraint in heavily phosphorus implanted silicon by HCl oxidation", Jpn. J. Appl. Phys., 18 (5), 873-80 (1979), EEA82-34305.

734. Hokari, Y. and Shiraki, H., "Secondary defect generation suppression in heavily boron implanted silicon wafers by HCl oxidation", Jpn. J. Appl. Phys., 16, (11), 1899-905 (1977), EEA81-5918.

735. Hu, S. M., "Hardening silicon wafers by ion implantation", IBM Tech. Disclosure Bull., 19 (2), 450-7 (1976).

736. Igras, E., "The effect of removal of gallium, thallium and indium acceptors and bismuth, antimony and arsenic donors in ion-bombarded silicon surface layers", Thin Solid Films, 34 (1), 124 (1976), PA79-62832.

737. Igras, E., "Anomalous surface diffusion of donors along radiation damaged regions in silicon", Proceedings of the International Conference on Ion Beam Modification of Materials, 1978, p. 371-80.

738. Kalinin, V. V., Aseev, A. L., Gerasimenko, N. N., Obodnikov, V. I. and Stenin, S. I., "The formation of defects in Si under the radiation enhanced diffusion conditions", Proceedings of the International Conference on Ion Beam Modification of Materials, 1978, p. 1161-76.

739. Kawado, S., "Control of oxidation-induced stacking faults in silicon by chlorine implantation", Jpn. J. Appl. Phys., 18 (2), 225-32 (1979), EEA82-21931.

740. Kool, W. H., Roosendaal, H. E., Wiggers, L. W. and Saris, F. W., "Damage production and arsenic displacement in silicon by proton and helium irradiation", Nucl. Instrum. & Methods, 132, 285-91 (1976), PA79-53188.

741. Koyama, R. Y., Phillips, W. E., Myers, D. R., Liu, Y. M. and Dietrich, H. B., "The energy levels and the defect signature of sulfur-implanted silicon by thermally stimulated measurements", Solid-State Electron, 21 (7), 953-5 (1978), EEA81-49616.

742. Lambert, J. A. and Dobson, P. S., "Defects in ion implanted silicon", Electron Microscopy 1978, Ninth International Congress on Electron Microscopy, p. 386-7, PA83-107300.

743. Mader, S. and Michel, A. E., "Residual lattice damage in As-implanted and annealed Si", J. Vac. Sci. & Technol., 13 (1), 391-5 (1976), PA79-58553.

744. Maksimov, S. K., Verner, V. D. and Piskunov, D. I., "Change in the mechanism for propagation of mismatch dislocations networks in silicon with a change of temperature", Sov. Phys.-Solid State, 17 (7), 1387-8 (1975), PA79-45090.

745. Miyao, M., Yoshihiro, N. and Tokuyama, T., "Damage dependent electrical activation of phosphorus implanted in silicon", Radiat. Eff., 47 (1-4), 85-90 (1980), EEA83-44009.

746. Miyao, M., Yoshihiro, N., Tokuyama, T. and Mitsuishi, T., "Damage-dependent electrical activation of ion-implanted silicon I. experiments on phosphorus implants", J. Appl. Phys., 50 (1), 223-30 (1979).

747. Narayan, J., "Formation of dislocation networks in arsenic implanted and thermally annealed silicon", Electrochemical Society Fall Meeting, Extended Abstracts, 1979, p. 1247-8.

748. Nylandsted Larsen, A., Weyer, G. and Nanver, L., "Radiation defects on ion-implanted silicon. II. Mossbauer spectroscopy of ^{119}Sn defect structures from implantations of radioactive tellurium", Phys. Rev. B, 21 (11), 4951-66 (1980), PA83-97175.
749. Orel, V., "Radiation defects due to Te^+ ion implantation in silicon", Scr. Fac. Sci. Nat. Univ. Purkynianae Brunensis Phys. 4 (2), 79-88 (1974), PA79-13541.
750. Orel, V., "Radiation defects in silicon due to ions N^+ and Te^+ implantation", Cesk. Cas. Fis. A, 25 (4), 376-9 (1975), Czech, PA79-26068.
751. Pavlidis, P., Economou, N. A. and Christou, A., "On the nature of dislocation loops in As implanted and postannealed Si wafers", J. Appl. Phys., 48 (8), 3612-13 (1977), EEA80-41223.
752. Pavlov, P. V., Pashkov, V. I., Tulovchikov, V. S. and Chigirinskaya, G. Y., "Influence of radiation damage on impurity diffusion from implanted layer in silicon", Sov. Phys.-Semicond., 11 (2), 181-3 (1977), EEA80-36239.
753. Pina, L., "Radiation damage profiles in $^1H^+$ and $^4He^+$ implanted Si crystals investigated by capacitance feedback method", Proceedings of the International Conference on Ion Beam Modification of Materials, 1978, p. 1673-9.
754. Rozgonyi, G. A. and Seidel, T. E., "Influence of wafer orientation and ion implantation on surface and bulk stacking faults", Electrochemical Society Spring Meeting, Extended Abstracts, 1977, p. 543-5, EEA81-5974.
755. Sadana, D. K., Washburn, J., Wilson, M. C. and Booker, G. R., "Regrowth behavior of three different damage structures in P^+ implanted and subsequently laser annealed Si", J. Electrochem. Soc., 127 (7), 1589-92 (1980).
756. Saraswat, K. C. and Meindl, J. D., "Epitaxial silicon growth on ion implanted silicon", Electrochemical Society Spring Meeting, Extended Abstracts, 1977, p. 311-12, EEA81-4981.
757. Schmitt, A. and Schorer, G., "Damage anneal of antimony/phosphorous double implants in silicon", Appl. Phys., 22 (2), 137-44 (1980).
758. Schmitt, A. and Schorer, G., "Method of eliminating lattice defects in ion implantation", IBM Tech. Disclosure Bull., 22 (10), 4601 (1980), EEA83-39751.
759. Schwuttke, G. H., Yang, K. and Kappert, H., "Lifetime control in silicon through impact sound stressing", Phys. Status Solidi A., 42 (2), 553-64 (1977).
760. Seshan, K. and Washburn, J., "Some new results in the characterization of defects in phosphorus ion-implanted silicon", Radiat. Eff., 26 (1-2), 31-7 (1975), EEA79-8782.
761. Seshan, K. and Washburn, J., "A three-stage model for the development of secondary defects in ion-implanted silicon", Radiat. Eff., 37 (3-4), 147-53 (1978), PA81-87725.
762. Sigmon, T. W., "Epitaxial regrowth of damage layers created by high dose ion implantation into Si", IEEE Trans. Electron Devices, ED-22 (11), 1065 (1975), EEA79-4546.

763. Tamura, M., "Dislocation networks in phosphorus-implanted silicon", Philos Mag., 35 (3), 663-691 (1977).
764. Tamura, M., Yagi, K., Natsuaki, N., Miyao, M. and Tokuyama, T., "Disorders produced during high-current and high-dose phosphorus ion implantation in silicon", Appl. Phys., 20 (3), 225-9 (1979).
765. Tamura, M., Yagi, K., Sakudo, N., Tokiguti, K., and Tokuyama, T., "High-current and high-dose phosphorus implantation in silicon", Proceedings of the International Conference on Ion Beam Modification of Materials, 1978, p. 515-24.
766. Tamura, M., Yoshihiro, N. and Ikeda, T., "Tertiary defects in phosphorus-implanted silicon", Appl. Phys. Lett., 27 (8), 427-9 (1975), PA79-9267.
767. Tamura, M., Yoshihiro, N. and Tokuyama, T., "Lattice defects and strains remaining in heavily ion-implanted silicon layers after high-temperature annealing", Electrochemical Society Spring Meeting, Extended Abstracts, 1977, p. 567-8, EEA81-5947.
768. Thompson, D. A. and Walker, R. S., "Low energy ion induced damage in silicon at 50K", Nucl. Instrum. & Methods, 132, 281-4 (1976), PA79-53187.
769. Wang, K. L., "Primary defects in low-fluence ion-implanted silicon", Appl. Phys. Lett., 36 (1), 48-50 (1980), PA83-44238.
770. Weyer, G., Nylandsted Larsen, A., Holm, N. E. and Nielsen, H. L., "Radiation defects in ion-implanted silicon. I. Mossbauer spectroscopy of ^{119}Sn defect structures from implantations of radioactive antimony", Phys. Rev. B, 21 (11), 4939-50 (1980), PA83-97174.
771. Wu, W. K. and Washburn, J., "On the shrinkage of rod-shaped defects in boron-ion implanted silicon", J. Appl. Phys., 48 (9), 3742-6 (1977), EEA81-5723.

j. Structure Analysis

772. Akimchenko, I. P., Kisselva, K. V., Krasnopevtsev, V. V., Touryansky, A. G. and Vavilov, V. S., "Structure and optical properties of silicon implanted by high doses of 70 and 310 keV carbon ions", Proceedings of the International Conference on Ion Beam Modifications of Materials, 1978, p. 1177-86.
773. Alessandrini, E. I., Chu, W. K. and Poponiak, M. R., "TEM study of the two-step annealing of arsenic-implanted <100> silicon", J. Vac. Sci. & Tech., 16 (2), 342-4 (1979), EEA82-42462.
774. Alstrup, O., "Investigation of radiation damage in ion implanted silicon crystals by pendellosung topography", Phys. Status Solidi A, 51 (2), 407-18 (1979), EEA82-26752.
775. Antonenko, A. K., Gerasimenko, N. N. and Dvurechenskii, A. V., "Electron spin resonance of conduction electrons in ion-implantation-doped silicon layers. Inhomogeneity of the impurity distribution", Sov. Phys.-Semicond., 11 (3), 322-4 (1977), PA83-2738.

776. Baither, D. and Bartsch, H., "TEM investigations of ion-implanted silicon", Proceedings of the International Conference on Ion Beam Modification of Materials, 1978, p. 1009-16.
777. Beanland, D. G., "The behaviour of boron molecular ion implants into silicon", Solid-State Electron, 21 (3), 537-47 (1978), EEA81-28424.
778. Beserman, R. and Bernstein, T., "Raman scattering measurements of the free-carrier concentration and of the impurity location in boron-implanted silicon", J. Appl. Phys., 48 (4), 1548-50 (1977), EEA80-24870.
779. Campisano, S. U., Rimini, E., Baeri, P. and Foti, G., "Supersaturated solid solutions after solid phase epitaxial growth in Bi-implanted silicon", App. Phys. Lett., 37 (2), 170-2 (1980), PA83-103384.
780. Christodoulides, C. E., Baragiola, R. A., Chivers, D., Grant, W. A. and Williams, J. S., "The recrystallization of ion-implanted silicon layers. II. Implant species effect", Radiat. Eff., 36 (1-2), 73-82 (1978), EEA81-28273.
781. Christodoulides, C. E., Grant, W. A. and Williams, J. S., "Dose dependence of residual lattice disorder in ion-implanted and annealed silicon", Appl. Phys. Lett., 30 (7), 322-3 (1977), EEA80-24696.
782. Comer, J. J., "Electron microscope study of stacking fault formation in boron implanted silicon", Radiat. Eff., 36 (1-2), 57-61 (1978), EEA81-28272.
783. Comer, J. J. and Roosild, S. A., "Transmission electron microscope study of oxidation-induced precipitate defects in boron-implanted silicon", J. Electron. Mater., 6 (3), 221-36 (1977), EEA81-19263.
784. Cullis, A. G., Webber, G. C., Poate, J. M. and Chew, N. G., "TEM study of silicon laser annealed after the implantation of low solubility dopants", J. Microsc., 118 (1), 41-9 (1980), EEA83-34616.
785. Damgaard, S., Petersen, J. W. and Holek, O., "Dose dependence of Mossbauer spectra of ion-implanted ^{57}Co and ^{57}Fe in silicon" Phys. Status Solidi A, 58 (2), 443-9 (1980), PA83-75900.
786. Drosd, R., "Transmission electron microscope study of complex annealing behavior of amorphous layers due to ion implantation in silicon", Electron Microscopy 1978. Ninth International Congress on Electron Microscopy, p. 384-5, PA83-107299.
787. Dvurechensky, A. V., Gerasimenko, N. N., and Glazman, V. B., "Defect annealing investigation in ion implanted Si by CESR technique", Radiat. Eff., 31 (1), 37-40 (1976).
788. Federovich, N. A., "Formation of complexes of arsenic implanted in silicon", Sov. Phys.-Solid State, 21 (8), 1413-14 (1979), PA83-59511.
789. Fletcher, J. and Booker, G. R., "TEM examinations of ion implanted silicon using cross-section specimens", Electron Microscopy 1978, Ninth International Congress on Electron Microscopy, p. 364-5, PA83-107298.

790. Gill, S. S. and Wilson, I. H., "Rutherford backscattering analysis of oxide layers formed by ion implantation into single-crystal silicon", Thin Solid Films, 55 (3), 435-48 (1978), CA90-178981.
791. Hinkel, H., Kaus, G. and Kempf, J., "Analysis of doped silicon by combined SIMS-RBS", IBM Tech. Disclosure Bull., 22 (3), 621-2 (1979), EEA83-9695.
792. Itoh, N., Morikawa, Y., Nakau, T. and Kuroda, H., "X-ray topographic studies of strains in silicon implanted with indium ions at high doses", Jpn. J. Appl. Phys., 14 (12), 2069-70 (1975), CA84-114340.
793. Joannou, D. E. and Davidson, S. M., "SEM-EBIC studies of boron implanted silicon", J. Microsc., 118, pt. 3, 337-42 (1980), EEA83-48573.
794. Josquin, W. J. M. J. and Tamminga, Y., "Rutherford backscattering study of crystal orientation dependent annealing effects in high-dose antimony implanted silicon", Appl. Phys., 15 (1), 73-8 (1978), EEA81-10262.
795. Komarov, F. F., Kuryazov, V. D., Solovev, V. S. and Shiryaev, S. T., "Formation of structure defects and behavior of ion-implanted boron in silicon under annealing in various ambients", Phys. Status Solidi A, 58 (1), 99-105 (1980), PA83-65033.
796. Korolev, N. I., Mironova, G. Y., Rembeza, S. I. and Spirin, A. I., "Study of ion-implanted silicon layers by IR- and EPR-spectroscope methods", Fiz.-Khimiya Poluprovodn. Materialoved., Voronezh, 41-5 (1978), Russian, CA91-31612.
797. Krimmel, E. F., Oppolzer, H. and Runge, H., "Transmission electron microscopical imaging of lateral implantation effects near mask edges in B^+-implanted Si wafers", Rev. Phys. Appl. 13 (12), 791-5 (1978), EEA82-17406.
798. Krynicki, J. and Bourgoin, J. C., "Defect annealing in phosphorus implanted silicon: a D.L.T.S. study", Appl. Phys., 18 (3), 257-8 (1979), EEA82-21922.
799. Kucirkova, A., "Ellipsometric study of silicon implanted with Li, N and Te ions", Scr. Fac. Sci. Nat. Univ. Purkynianae Brunensis Phys., 4 (2), 59-64 (1974), EEA79-8798.
800. Kucirkova, A., "Ellipsometric study of tellurium implanted silicon", Radiat. Eff., 28 (3-4), 129-31 (1976), PA79-62373.
801. Lambert, J. A. and Dobson, P. S., "The structure of rod defects in boron-implanted silicon", Philos. Mag. [Part] A, 37 (4), 441-6 (1978), CA89-189092.
802. Larson, B. C. and Barhorst, J. F., "X-ray study of lattice strain in boron implanted laser annealed silicon", J. Appl. Phys., 51 (6), 3181-5 (1980).
803. Larson, B. C., White, C. W. and Appleton, B. R., "Unidirectional contraction in boron-implanted laser-annealed silicon", Appl. Phys. Lett., 32 (12), 801-3 (1978), EEA81-42340.

804. Lau, S. S., Matteson, S., Mayer, J. W., Revesz, P., Gyulai, J., Roth, J., Sigmon, T. W. and Cass, T., "Improvement of crystalline quality of epitaxial Si layers by ion-implantation techniques", App. Phys. Lett., 34 (1), 76-8 (1979), EEA82-21214.
805. Lecrosnier, D. P. and Pelous, G. P., "Distribution of secondary defects in boron-implanted silicon", Lattice Defects in Semiconductors, 1974, p. 487-93, Pub. 1975, PA78-81132.
806. Lecrosnier, D. P., Pelous, G. P. and Burgeat, J., "Investigation of ion-implantation damage and x-ray double reflection", Appl. Phys. Lett., 30 (3), 141-3 (1977), EEA80-9393.
807. Ling, H., "Electrical and electron microscopy observations on defects in ion implanted silicon", Report LBL-7386, California Univ., Berkeley, 1978, 28 pp.
808. Mader, S. and Tu, K. N., "Microstructure of xenon-implanted silicon", J. Vac. Sci. & Technol., 12 (1), 501-3 (1975), PA79-9320.
809. Maksimov, S. K., Lukyanchuk, T. I. and Piskunov, D. I., "Two mechanisms of the formation of a defect structure in silicon in the case of ion implantation of impurities and subsequent annealing", Sov. Phys.-Dokl., 20 (8), 577-9 (1975), PA79-44920.
810. Marcyk, G. T. and Streetman, B. G., "Glow discharge optical spectroscopy measurements of arsenic-implanted silicon", J. Vac. Sci. & Technol., 14 (5), 1165-7 (1977), EEA81-14655.
811. Mayer, J. W., Csepregi, L., Gyulai, J., Nagy, T., Mezey, ., Revesz, P. and Kotai, E., "MeV helium backscattering analysis of ion-implanted silicon drive-in diffusion and epitaxial regrowth", Thin Solid Films, 32 (2), 303-6 (1976), CA84-158747.
812. McMahon, R. A. and Ahmed, H., "Surface structure studies of electron beam annealed ion implanted silicon", J. Microsc., 118 (1), 61-6 (1980), EEA83-34618.
813. Miyao, M., Miyazaki, T. and Tokuyama, T., "Optical reflectivity studies of damage in ion implanted silicon", Jpn. J. Appl. Phys 17 (5), 955-6 (1978), EEA81-33271.
814. Motamedi, M. E. and Das, P., "Study of ion implanted silicon surface using acoustoelectric voltage", 1976 Ultrasonics Symposium Proceedings, 1976, p. 205-9, EEA80-13366.
815. Muller, J. C., "Study of the defects introduced by boron ion implantation in silicon", Report CRN-PN-76-2, Cent. Rec. Nucl., Univ. Strasbourg, Strasbourg, France, 1976, 83 pp., French, CA85-185620.
816. Nakamura, K., Gotoh, T. and Kamoshida, M., "Characterization of $^{31}P^{+}$-implanted Si layers by ellipsometry", J. Appl. Phys., 50 (6), 3985-9 (1979), EEA82-46904.
817. Nakamura, K. and Kamoshida, M., "Characterization of laser-annealed Si layers by ellipsometry", Radiat. Eff., 42 (1-2), 29-34 (1979), EEA83-19000.
818. Nakamura, K., Kamoshida, M., Uchara, A. and Tatsumi, R., "Ellipsometric analysis of laser-annealed Si layers", AIP Conf. Proc. no. 50, 434-9 (1978), EEA82-38230.

819. Nangia, O. P., "Qualitative check of phosphorus distribution in neutron irradiated silicon crystals using autoradiography", Curr. Sci., 48 (15), 678-9 (1979), EEA83-14280.
820. Narayan, J., White, C. W. and Young, R. T., "Transmission electron microscope study of laser annealing of ion damage in silicon and gallium arsenide", Proceedings of the International Conference on Ion Beam Modification of Materials, 1978, p. 819-30.
821. Ohdomari, I., Ikeda, M., Ohna, K. and Itoh, T., "Structure of surface layers of silicon heavily implanted with Ar", J. Appl. Phys., 46 (12), 5065-8 (1975), EEA79-16607.
822. Orel, V. and Ohlidal, M., "Preparation of ion-implanted silicon thin foils for transmission electron microscopy", Scr. Fac. Sci. Nat. Univ. Purkynianae Brunensis Phys., 4 (2), 89-96 (1974), PA79-14720.
823. Perezhogin, G. A., "Effects of isotopic-composition shifts on neutron-activation determination of ion-implanted antimony", Ind. Lab., 44 (10), 1389-90 (1978), PA83-24487.
824. Picraux, S. T. and Vook, F. L., "Structure of hydrogen center in D-implanted Si", Phys. Rev. B, 18 (5), 2066-77 (1978), EEA82-4177.
825. Porter, W. A., Parker, D. L. and Richardson, T. W., "Characterization of laser irradiated silicon using X-ray topography", AIP Conf. Proc., no. 50, 453-6 (1978), EEA82-38233.
826. Prasad, A., Beserman, R., Germain, P. and Bourgoin, J. C., "Kinetics of recrystallization in ion-implanted silicon layers", Phys. Status Solidi A, 35 (1), 109-112 (1976).
827. Richon, F., Pelous, G. and Lecrosnier, D., "Selenium implantation into silicon studied by the DLTS technique", Appl. Phys. Lett., 31 (8), 525-7 (1977), EEA81-5721.
828. Romanov, S. I. and Smirnov, L. S., "Voids in ion-implanted silicon", Radiat. Eff., 37 (1-2), 121-6 (1978), PA81-67757.
829. Runovc, F., "Effectiveness of charged vacancies in diffusion of implanted boron in silicon", Electron Lett., 16 (2), 49-50 (1980), EEA83-14282.
830. Sadana, D. K., Fletcher, J. and Booker, G. R., "Structural and electrical profiles for double damage layers in ion-implanted silicon", Electron. Lett., 13 (21), 632-3 (1977), EEA81-1019.
831. Sadana, D. K., Wilson, M. C. and Booker, G. R., "Electron microscope studies of ion implanted silicon and gallium arsenide after laser and furnace annealing", J. Microsc. 118, (pt. 1), 51-9 (1980), EEA83-34617.
832. Schartner, K. H., Hoogkamer, T. P., Woerlee, P. and Saris, F. W., "Ion-implantation method to measure the fraction of beam carrying inner-shell vacancies when penetrating solids", Nucl. Instrum. Methods, 132, 35-38 (1976).
833. Schoen, N. C., "Thermoelastic stress analysis of pulsed electron beam recrystallization of ion-implanted silicon", J. Appl. Phys., 51 (9), 4757-69 (1980), PA84-7251.

834. Schirmer, G., Jahn, U. and Frey, H., "Investigation of boron implanted silicon crystals", Wiss. Z. Friedrich-Schiller-Univ. Jena. Math. Naturwiss. Reihe, 27 (2-3), 199-211 (1978), German, EEA81-45945.
835. Schulson, E. M. and Marsden, D. A., "A rapid assessment of mercury ion implantation damage in boron-doped silicon using the scanning electron microscope", Proc., Annu. Conf. - Microbeam Anal. Soc., 1974, 6 pp., CA87-110333.
836. Seidel, T. E., Pasteur, G. A. and Tsai, J. C. C., "Visible interference effects in silicon caused by high-current high-dose implantations", Appl. Phys. Lett., 29 (10), 648-51 (1976), EEA80-6480.
837. Seshan, K., "Electron microscope studies of ion implanted silicon", Thesis, University of California at Berkeley, Berkeley, CA, 1975, 108 pp., Order No. 76-15364.
838. Seshan, K., Observations on the development of secondary defects in heavily damaged ion-implanted (100) and (111) silicon", Radiat. Eff., 40 (1-2), 29-35 (1979), PA82-51166.
839. Shevlin, C. M. and Demer, L. J., "Rapid indentification of faulted loops in ion-implanted silicon", Radiat. Eff., 38 (3-4), 197-9 (1978), PA82-19254.
840. Sorokin, L. M. and Sitnikova, A. A., "Isolated disturbed zones in boron-ion-doped silicon layers", Sov. Phys.-Solid State, 20 (2), 290-2 (1978), EEA81-49733.
841. Stelmakh, V. F., Tkachev, V. D. and Chelyadinskii, A. R., "X-ray diffraction investigation of silicon doped by boron ion implantation", Sov. Phys.-Solid State, 20 (7), 1267-70 (1978), PA82-37726.
842. Stetter, G., Coufal, H. and Luscher, E., "EPR study of the annealing behaviour of neutron-transmutation-doped silicon", Phys. Status Solidi A, 49 (1), K17-20 (1978), EEA82-811.
843. Sugiura, H., "Study of method combined with silicon vacuum deposition and antimony ion implantation", Electr. Commun. Lab. Tech. J., 27 (6), 1329-36 (1978), EEA82-12563.
844. Suski, J. and Rzewuski, H., "Ionization-enhanced crystallization of phosphorus implanted silicon layers", Radiat. Eff., 40 (1-2), 81-6 (1979).
845. Tamura, M., Yagi, K., Natsuaki, N., Miyao, M. and Tokuyama, T., "Disorders produced during high-current and high-dose phosphorus ion implantation in silicon", Appl. Phys., 20 (3), 225-9 (1979), PA83-6422.
846. Tyagai, V. A., Lysenko, V. S., Krasiko, A. N., Evstigneev, A. M. and Lokshin, M. M., "Anisotropic-electroreflectance determination of surface lattice defects in ion implantation", Sov. Tech. Phys. Lett., 4 (1), 39-40 (1978), PA82-19274.
847. Vale, R. J., "Ion implantation of silicon", Electron Microscopy Society of Southern Africa 15th Annual Conference, 1976, p. 89-90, EEA80-24883.

848. Watanabe, K., Miyao, M., Takemoto, I. and Hashimoto, N., "Ellipsometric study of silicon implanted with boron ions in low doses", Appl. Phys. Lett., 34 (8), 518-19 (1979), EEA82-34262.
849. Weyer, G., Damgaard, S., Peterson, J. W. and Heinemeier, J., "Mossbauer study of ^{119}Sn defects in silicon from ion implantations of radioactive ^{119}In", Hyperfine Interactions, 7 (6), 449-53 (1980), PA83-80641.
850. Weyer, G., Kettschau, A., Grebe, G., Schroter, W. and Bergholz, W., "Comparison of Mossbauer spectra of ^{57}Co in silicon for high-temperature ion-implanted and for diffused samples", Phys. Status Solidi A, 51 (2), 459-66 (1979), PA82-51891.
851. Williams, J. S., Christodoulides, C. E. and Grant, W. A., "The influence of high implant concentration on solid phase epitaxial regrowth in (100) and (111) silicon", Proceedings of the International Conference on Ion Beam Modification of Materials, 1978, p. 969-76.
852. Williams, J. S., Christodoulides, C. E., Grant, W. A., Andrew, R., Brawn, J. R. and Booth, M., "The recrystallization of ion implanted silicon layers. I. Pb-ion implanted Si", Radiat. Eff., 32 (1-2), 55-66 (1977), EEA80-24879.
853. Williams, J. S. and Grant, W. A., "The application of high resolution Rutherford backscattering to the measurement of ion ranges in Si and Al", Radiat. Eff., 25 (1), 55-6 (1975), PA79-5448.
854. Wu, W. K. and Washburn, J., "Rod-like defects in ion implanted silicon", Cryst. Lattice Defects, 7 (1), 39-43 (1977), PA80-68975.
855. Wu, W. K. and Washburn, J., "On the climb on dislocations in boron-ion implanted silicon", J. Appl. Phys., 48 (9), 3747-51 (1977), PA81-6062.
856. Zielinska-Rohozinska, E., "X-ray diffraction investigations of annealing characteristics in P^+-implanted silicon crystals", Phys. Status Solidi A, 44 (1), 59-64 (1977), EEA81-10121.

k. Through SiO_2

857. Hirao, T., Fuse, G., Inoue, K., Takayanagi, S., Yaegashi, Y. and Ichikawa, S., "Effects of the recoil-implanted oxygen in Si on the electrical activation of As after through-oxide implantation", J. Appl. Phys., 50 (8), 5251-6 (1979).
858. Hirao, T., Fuse, G., Inoue, K., Takayanagi, S., Yaegashi, Y., Ichikawa, S. and Izumi, T., "Electrical properties of Si implanted with As through SiO_2 films", J. Appl. Phys., 51 (1), 262-8 (1980).
859. Kawata, H., Sato, T. and Sakurai, T., "Lateral spread of B^+ ions implanted in a silicon substrate through a SiO_2 mask", Fujitsu Sci. & Tech. J., 14 (4), 83-93 (1978), EEA82-26748.

860. Kotai, E., Nagy, T., Meyer, O., Gyulai, J., Revesz, P., Mezey, G., Lohner, T. and Manuaba, A., "Diffusion measurement of implanted Sb into Si, using SiO_2 encapsulation", Proceedings of the International Conference on Ion Beam Modification of Materials, 1978, p. 573-82.
861. Kubota, H., Hayashi, Y., Okazaki, N. and Watanabe, S., "Ion implantation in semiconductor substrates", Patent Germany 2,649,134, Publ. May 1977, CA87-15022.
862. Myers, D. R., "Properties of silicon implanted with arsenic through silicon dioxide", Thesis, 1977, Order No. 77-26718, CA88-56687.
863. Natsuaki, N., Tamura, M., Miyao, M. and Tokuyama, T., "Anomalous residual defects in silicon after annealing of through-oxide phosphorus implanted samples", Jpn. J. Appl. Phys., 16, suppl. 1, 47-51 (1977), EEA80-36232.
864. Natsuaki, N., Tamura, M. and Tokuyama, T., "Low energy and high dose phosphorus ion implantation into silicon through SiO_2 film", Jap. J. Appl. Phys., 15 (12), 2427-32 (1976), EEA80-9402.
865. Okabayashi, H., Yoshida, M., Ishida, K. and Yamane, T., "SiO_2 antireflection coating effect on cw laser annealing of ion implanted Si", Appl. Phys. Lett., 36 (3), 202-3 (1980).
866. Panknin, D., Ross, R. and Mende, G., "Effect of silicon dioxide layers on profiles in phosphorus-implanted and annealed silicon", Zentralinst., Kernforsch., Rossendorf Dresden, ZfK-350, 156-8 (1978), German, CA89-207937.
867. Sakurai, T., Kawata, H., Sato, T., Hisatsugu, T., Hashimoto, H. and Furuya, T., "Lateral spread of P^+ ions implanted in silicon through the SiO_2 mask window", J. Appl. Phys., 50 (3), 1287-90 (1979), EEA82-38251.
868. Tamura, H., Miyao, M. and Tokuyama, T., "Laser-annealing behaviour of a phosphorus-implanted silicon substrate covered with a SiO_2 film", J. Appl. Phys., 50 (5), 3783-4 (1979), EEA82-42457.

l. Others

869. Akasaka, Y. and Tsukamoto, K., "Arsenic implantation to silicon", Oyo Buturi, 46 (11), 1131-7 (1977), Japanese, EEA81-23872.
870. Beanland, D. G., "The behaviour of boron molecular ion implants into silicon", Solid-State Electron., 21 (3), 537-47 (1978), EEA81-28424.
871. Bentini, G. G., Lotti, R., Nipoti, R. and Nobili, D., "Ion implanted induced giant gettering of gold in silicon", Radiat. Eff. Lett., 50 (3-6), 153-9 (1980), PA83-88221.
872. Beyer, K. D. and Yeh, T., "Ion-implantation gettering at elevated temperatures", IBM Tech. Disclosure Bull., 20 (8), 3122 (1978), EEA82-4299.

873. Bogrdus, E. H., "Ion-implantation into As-diffused junctions in Si", J. Appl. Phys., 48 (5), 2093-4 (1977), EEA80-28777.

874. Bullard, R. L. and Hobbs, C. W., "Delineation of ion-implanted semiconductor devices for physical measurement", IBM Tech. Disclosure Bull., 21 (8), 3277 (1979), EEA82-38249.

875. Druzhkov, A. P., Khaibullin, I. B., Zaripov, M. M. and Bayazitov, R. M., "Positron annihilation study of implanted silicon layers", Int. Conf. Ion Implantation Semicond., Proc., 1978, p. 504-14, CA90-131609.

876. Fair, R. B. and Pappas, P. N., "The gettering of boron by an ion-implanted antimony layer in silicon", Solid-State Electron., 18 (12), 1131-4 (1975), PA79-10492.

877. Geipel, H. J. and Tice, W. K., "Critical microstructure for ion-implantation gettering effects in silicon", Appl. Phys. Lett., 30 (7), 325-7 (1977), EEA80-24861.

878. Gotzlich, J. F., Haberger, K., Ryssel, H., Kranz, H. and Traumuller, E., "Dopant dependence of the oxidation rate of ion implanted silicon", Proceedings of the International Conference on Ion Beam Modification of Materials, 1978, p. 1419-32.

879. Grotzschel, R., Klabes, R., Kreissig, U. and Schmidt, A., "Recoil implantation from thin surface films on silicon", Radiat. Eff., 36 (3-4), 129-34 (1978), EEA81-38126.

880. Haas, E., Martin, J. and Schnoller, M., "Doping of silicon by nuclear transmutation", Siemens Rev., 44 (2), 90-3 (1977), EEA80-41397.

881. Hoffmeister, W., Schmitt, A. and Schumacher, H., "Method of reducing the arsenic doping in silicon", IBM Tech. Disclosure Bull., 22 (8A), 3255 (1980), EEA83-39746.

882. Hofker, W. K., "Implantation of boron in silicon", Thesis, Eindhoven Univ. Technol., Netherlands, 1975, PA79-27044.

883. Huller, J. and Pham, M. T., "Electrochemical characterization of ion implanted silicon", Proceedings of the International Conference on Ion Beam Modification of Materials, 1978, p. 1709-21.

884. Jaskolska, H., Walis, L. and Golkowska, H., "The need to use appropriate calibration curves for the anodic oxidation of ion-implanted silicon", Thin Solid Films, 33 (3), 281-6 (1976), CA85-12871.

885. Josquin, W. J. M. J., "The oxidation characteristics of nitrogen-implanted silicon", Proceedings of the International Conference on Ion Beam Modification of Materials, 1978, p. 1433-44.

886. Kiselevich, M., Lyatushinski, A., Zhuk, V. and Osipenko, B. P., "Implantation of heavy ions into silicon single crystals", Sov. Phys.-Solid State, 17 (4), 687-9 (1975), PA79-14717.

887. Kutukova, O. G. and Streltsov, L. N., "Influence of the radiation dose on diffusion processes in ion-implantation-doped layers", Sov. Phys.-Semicond., 10 (4), 475-6 (1976), EEA80-621.

888. Kuznetsov, V. P., Tolomasov, V. A. and Tumanova, A. N., "Doping of thin epitaxial Si layers grown in vacuum at low temperatures", Sov. Phys.-Crystallogr., 24 (5), 588-90 (1979), EEA83-44018.
889. Ligeon, E. and Guivarch, A., "Hydrogen implantation in silicon between 1.5 and 60 keV", Radiat. Eff., 27 (3-4), 129-37 (1976), PA79-40103.
890. Matsumoto, H. and Sugano, T., "Impurity doping of silicon by thermal neutron irradiation", Annu. Rep. Eng. Res. Inst. Fac. Eng. Univ. Tokyo, 38, 141-6 (1979), EEA83-23907.
891. Mende, G. and Kuster, G., "On the removal of thin layers of P- and B-implanted silicon by means of anodic oxidation", Thin Solid Films, 35 (2), 215-20 (1976), German, PA79-73313.
892. Miller, R. J., "PtSi or Pd_2Si self-aligned ion implantation mask over silicon", IBM Tech. Disclosure Bull., 23 (3), 1030-3 (1980).
893. Motamedi, M. E., Das, P. and Bharat, R., "Surface study of ion-implanted silicon wafers using SAW convolver", J. Appl. Phys., 48 (6), 2135-8 (1977), EEA80-36223.
894. Ota, Y., "n-type doping techniques in silicon molecular beam epitaxy by simultaneous arsenic ion implantation and by antimony evaporation", J. Electrochem. Soc., 126 (10), 1761-5 (1979), EEA83-9698.
895. Ota, Y., "Silicon molecular beam epitaxy with simultaneous ion implant doping", J. Appl. Phys., 51 (2), 1102-10 (1980), EEA83-28033.
896. Parry, P. D., Cho, K. H. and Bristol, S. P., "Silicon shadow masks for ion implantation", Electrochemical Society Spring Meeting, 1977, p. 285-7, EEA81-5933.
897. Possin, G. E. and Kirkpatrick, C. G., "Spatial variation of lifetime distributions due to inert ion damage from low and room temperature implantation", J. Vac. Sci. & Technol., 16 (6), 1917-20 (1979), EEA83-34808.
898. Ross, R., "The chemical behavior of high dose implanted silicon", Proceedings of the International Conference on Ion Beam Modification of Materials, 1978, p. 1445-50.
899. Runge, H. and Worl, H., "Autoradiographic detection of getter effect of argon-implanted layers in silicon", Phys. Status Solidi A, 45 (2), 509-12 (1978), EEA81-23873.
900. Rybinski, J., "Piezoresistance tensor coefficients of phosphorus implanted silicon thin films", Electron. Technol., 12 (2), 59-65 (1979), EEA83-48568.
901. Shannon, J. M., "Recoil-implanted antimony-doped surface layers in silicon", Applications of Ion Beams to Materials, 1975, p. 37-43, Publ. 1976, PA79-69555.
902. Shinada, K. and Itoh, T., "Boron implantation into Si using B_2O_3 and B mixture for ion source material", Jpn. J. Appl. Phys., 17 (10), 1839-44 (1978), EEA82-804.

903. Sigmon, T. W., Csepregi, L. and Mayer, J. W., "Ion implantation gettering of gold in silicon", J. Electrochem. Soc., 123 (7), 1116-17 (1976), EEA79-41384.
904. Sugiura, H., "A method of Si vacuum deposition combined with Sb ion implantation", J. Vac. Soc. Jpn., 21 (3), 80-5 (1978), Japanese, EEA81-49006.
905. Sugiyama, T., Shimada, O. and Itoh, T., "On the doping efficiency of Ga or Sb into silicon epitaxial films grown by part ionized vapor deposition", J. Vac. Soc. Jpn., 21 (11), p. 386-91 (1978), Japanese, EEA82-38255.
906. Tsai, M. Y., Streetman, B. G., Blattner, R. J. and Evans, C. A., Jr., "Study of surface contamination produced during high dose ion implantation", J. Electrochem. Soc., 126 (1), 98-102 (1979).
907. Tsukamoto, K., Akasaka, Y. and Kijima, K., "Thermal diffusion of ion-implanted arsenic in silicon", Jpn. J. Appl. Phys., 19 (1), 87-95 (1980).
908. Wagh, A. G., Sarma, N. and Bhattacharya, P. K., "Gallium-implanted silicon", Phys. Status Solidi A, 32 (1), 63-8 (1975), PA79-18902.
909. Williams, J. S., "The significance of ion implantation induced stress in silicon", Phys. Lett. A, 60A (4), 330-2 (1977), PA80-35610.
910. Yamaguchi, M. and Hiroyama, T., "Surface contamination of silicon produced by ion implantation", Jpn. J. Appl. Phys., 15 (2), 365-72 (1976).

B. Others

911. Akimchenko, I. P., Kisseleva, K. V., Krasnopevtsev, V. V., Milyutin, Y. V., Touryansky, A. G. and Vavilov, V. S., "The structure of silicon carbide synthesized in diamond and silicon by ion implantation", Radiat. Eff., 33 (2), 75-80 (1977), PA80-76434.
912. Akimchenko, I. P., Krasnopevtsev, V. V. and Milyutin, Yu. V., "Role of structural imperfections of the substrate in the synthesis of silicon carbide in the case of ion implantation in natural diamond", Sov. Phys. - Lebedev Inst. Rep., 37-40 (1977), EEA81-18677.
913. Akimchenko, I., Zavetova, M. and Zemek, J., "Influence of the ion-implanted oxygen on the absorption edge and electrical properties of amorphous Ge", Czech. J. Phys. B, B26 (12), 1409-10 (1976), EEA80-17096.
914. Andersen, H. H., Nathan, M. I. and Ziegler, J. F., "Storage of information by heavy ion irradiation of Si or Ge", IBM Tech. Disclosure Bull., 20 (9), 3772-3 (1978), EEA82-878.

915. Apsley, N., Davis, E. A., Yoffe, A. D. and Troup, A. P., "Electronic properties of ion-bombarded evaporated germanium and silicon", J. Phys. C, 11 (24), 4983-96 (1978), EEA82-12474.
916. Axmann, A., Schulz, M. and Fritzsche, C. R., "Implantation doping of germanium with Sb, As, and P", Appl. Phys., 12 (2), 173-8 (1977), EEA80-9394.
917. Bertolotti, M. and Vitali, G., "Regrowth of Si and Ge under laser irradiation", International Conference on Defects and Radiation Effects in Semiconductors, 1978, p. 454-9, EEA83-14308.
918. Boolchand, P., Van Rossum, M., Langouche, G., Odeurs, J., Parryn, H. and Coussement, R., "Mossbauer effect studies following ion implantation of ^{125}I and ^{125m}Te radioactivities in crystalline Ge", Hyperfine Interactions, 2 (2), 371-3 (1976), PA80-20758.
919. Braunstein, G., Bernstein, T., Carsenty, U. and Kalish, R., "Depth profile of antimony implanted into diamond", J. Appl. Phys., 50 (9), 5731-5 (1979), EEA83-744.
920. Braunstein, G., Talmi, A., Kalish, R., Bernstein, T. and Beserman, R., "Radiation damage and annealing in Sb implanted diamond", Radiat. Eff., 48 (1-4), 139-44 (1980), EEA83-53662.
921. Campisano, S. U., Grimaldi, M. G., Baeri, P., Foti, G. and Rimini, E., "Laser induced reorder in Pb implanted Ge", Appl. Phys., 22 (2), 201-4 (1980).
922. Della Mea, G., Foti, G. and Majni, G., "Laser-induced regrowth of (100) Ge implanted with B and P ions", AIP Conf. Proc., no. 50, 317-20 (1978), EEA82-34277.
923. El-Atawy, S. A. and Ade, P. A. R., "Far-infrared composite bolometers using surface ion implanted Ge as an absorbing surface", Infrared Phys., 18, (5-6), 683-90 (1978), EEA82-19477.
924. El-Atawy, S. A., Clegg, P. E. and Palmer, D. W., "Ion implantation and depletion in germanium at low temperatures", Infrared Phys., 19 (2), 185-9 (1979), EEA82-30372.
925. Evans, S., "Some effects of heteroatom size and reactivity on low-energy ion implantation in diamond, revealed by x-ray induced electron spectroscopy and oxidative depth profiling", Proc. R. Soc. London Ser. A, 370 (1740), 107-29 (1980), PA83-44226.
926. Foti, G., Della Mea, G., Jannitti, E. and Majni, G., "Laser reordering of implanted amorphous layer in germanium", Phys. Lett. Vol. A, 68A (3-4), 368-70 (1978), EEA82-8117.
927. Foti, G., Vitali, G. and Davies, J. A., "Molecular and atomic damage in germanium", Radiat. Eff., 32 (3-4), 187-91 (1977), EEA80-28833.
928. Glazman, V. B., Safronov, L. N. and Smirnov, L. S., "Influence of dislocations in germanium on the depth of a defect layer resulting from ion implantation", Sov. Phys.-Semicond., 9 (11), 1444-5 (1975), PA79-66141.

929. Gorkun, Y. I., Korobanov, V. L., Lysenko, V. S., Novominskii, V. A. and Turchanikov, V. I., "Effect of bombardment with low-energy Ar^+ ions on electron scattering in accumulation layer of n-Ge", Phys. Status Solidi A, 41 (1), 339-46 (1977), EEA80-32304.

930. Graczyk, J. F. and Chaudhari, P., "Effects of ion implantation on the structure of amorphous germanium", Appl. Phys. Lett., 32 (8), 466-8 (1978), PA81-55881.

931. Gusev, V. M., Demakov, K. D., Kosaganova, M. G., Reifman, M. B. and Stolyarova, V. G., "Investigation of the electroluminescence of α-SiC crystals implantation-doped with boron, aluminium, and gallium", Sov. Phys.-Semicond., 9 (7), 820-2 (1975), EEA79-20798.

932. Guseva, M. I., Konorova, E. A., Kuznetsov, Y. A. and Sergienko, V. F., "Double injection of carriers in a p-i-n structure made of implantation-doped diamond", Sov. Phys.-Semicond., 12 (3), 290-3 (1978), EEA82-8210.

933. Hubler, G. K., Malmberg, P. R. and Comas, J., "1.2 MeV aluminum ion implantation into silicon carbide", Rep. NRL Prog., 37-40 (1975), CA84-158732.

934. Ichikawa, T. and Ino, S., "Recovery processes of Si(111) and Ge(111) surfaces damaged by argon ion bombardment", Jpn. J. Appl. Phys., 17 (9), 1675-6 (1978), EEA81-49741.

935. Igras, E., "Light and radiation induced annealing effects in silicon and germanium implanted layers", Proceedings of the International Conference on Ion Beam Modification of Materials, 1978, p. 381-90.

936. Kalish, R., Deicher, M., Rechnagel, E. and Wichert, T., "Intrinsic limitations of doping diamonds by heavy-ion implantation", J. Appl. Phys., 50 (11), Pt. 1, 6870-2 (1979), PA83-34019.

937. Kamada, K., Kazumata, Y. and Kubo, K., "Observation of blistering and amorphization on germanium surface after 450 keV Ar^+ ion bombardment", Radiat. Eff., 28 (1-2), 43-8 (1976), PA79-53195.

938. Kodrau, N. V. and Makarov, V. V., "Investigation of the blue luminescence of silicon carbide doped by ion implantation of aluminium and nitrogen", Sov. Phys.-Semicond., 11 (5), 572-3 (1977), EEA81-10643.

939. Konorova, E. A., Sergienko, V. F., Tkachenko, S. D., Dravin, V. A. and Spitsyn, A. V., "Investigation of the distribution of ion-implanted lithium atoms in diamond", Sov. Phys.-Lebedev Inst. Rep., no. 4, 9-14 (1978), EEA82-38260.

940. Kryuchkov, Y. Y., Timoshnikov, Y. A. and Chernow, I. P. "Determination of the localization of oxygen in the germanium lattice after large implantation doses", Sov. Phys.-Semicond., 13 (2), 218-19 (1979), EEA82-47087.

941. Lee, Y. H., Brosious, P. R. and Corbett, J. W., "High-temperature ion implantation in diamond", Phys. Status Solidi A, 50 (1), 237-42 (1978), EEA82-12573.

942. Makarov, V. V., Tuomi, T., Naukkarinen, K., Luomajarvi, M. and Riihonen, M., "Laser-induced recrystallization and defects in ion-implanted hexagonal SiC.", Appl. Phys. Lett., 35 (12), 922-24 (1979), PA83-44221.
943. McDonald, R. E. and McNab, T. K., "Lattice location of ^{12}B in germanium and silicon at 300K", Phys. Rev. B, 13 (1), 39-41 (1976), PA79-40076.
944. Mitchell, J. B., Foti, G., Howe, L. M. and Davies, J. A., "A channeling investigation of proton and deuteron damage in germanium", Radiat. Eff., 26 (3), 193-9 (1975), PA79-13539.
945. Novozhilov, O. A., Novikov, S. R., Khaibullin, I. B., Shtyrkov, E. I., Zakirov, G. G. and Romanov, S. I., "Influence of irradiation and annealing regimes on the optical properties and structure of implanted germanium layers", Proceedings of the International Conference on Ion Beam Modification of Materials, 1978, p. 1341-54.
946. Pavlov, P. V., Zorin, E. I., Tetelbaum, D. I. and Khokhlov, A. F., "Nitrogen as dopant in silicon and germanium", Phys. Status Solidi A, 35 (1), 11-36 (1976), EEA79-32888.
947. Rahn, L. A., Colwell, P. J., Choyke, W. J., Urli, N. B. and Corbett, J. W., "Raman scattering from ion-implanted silicon carbide", International Conference on Radiation Effects in Semiconductors, 1976, p. 515-16, PA81-6689.
948. Sekhar, P., Joshi, M. C., Narasimhan, K. L. and Guha, S., "Ion implantation in amorphous germanium and silicon", Solid State Commun., 26 (12), 933-6 (1978), EEA81-45943.
949. Shtyrkov, E. I., Khaibullin, I. B., Zaripov, M. M., Galyatudinov, M. F. and Bayazitov, R. M., "Local laser annealing of implantation doped semiconductor layers", Sov. Phys.-Semicond., 9 (10), 1309-10 (1975), PA79-58562.
950. Sobczak, E., "X-ray bremsstrahlung isochromat and x-ray photoelectron investigations of ion implanted germanium", Phys. Status Solidi A, 54 (2), K151-3, (1979), PA83-2360.
951. Soeder, J. M., "Experiments for the preparation of Nb_3Sn, Nb_3Ge and Nb_3Si by ion implantation", Report Juel-1474, Kernforschungsanlage Juelich G.m.b.H., Germany, F. R., 1977, 58 pp.
952. Stas, V. F. and Smirnov, L. S., "Radiation enhanced P-diffusion in germanium", Proceedings of the International Conference on Ion Beam Modification of Materials, 1978, p. 497-504.
953. Thompson, D. A., Chan, M. C. and Campbell, A. B., "Nitrogen implanted SiC: correlation of channeling and electrical studies", Can. J. Phys., 54 (6), 626-32 (1976), EEA79-24703.
954. Thompson, D. A. and Walker, R. S., "Energy spikes in Si and Ge due to heavy ion bombardment", Radiat. Eff., 36 (1-2), 91-100 (1978), EEA81-28274.
955. Thompson, D. A., Walker, R. S. and Davies, J. A., "Evidence for spike-effects in low-energy heavy-ion bombardment of Si and Ge", Radiat. Eff., 32 (3-4), 135-42 (1977), EEA80-28831.

956. Tsaur, B. Y., Liau, Z. L. and Mayer, J. W., "Formation of silicon enriched metastable compounds in the platinum-silicon system using ion bombardment and post annealing", Phys. Lett. A, 71A (2-3), 270-2 (1979), CA91-67261.
957. Vavilov, V. S., "Ion implantation into diamond", Radiat. Eff., 37 (3-4), 229-36 (1978), EEA81-45941.
958. Weyer, G., Damgaard, S., Petersen, J. W. and Heinemeier, J., "Sn impurity defects in germanium from ion implantation of radioactive ^{119}In", Phys. Lett. A, 76A (3-4), 321-3 (1980), PA83-59495.
959. Zellama, K., Germain, P., Bourgoin, J. C. and Morhange, J. F., "Growth rate of crystallization in amorphous germanium produced by ion implantation. A Raman spectroscopy study", Phys. Status Solidi A, 56 (2), 717-23 (1979), PA83-44227.
960. Zemek, J., Zavetova, M., Akimchenko, I. and Milyutin, Y. V., "Influence of ion-implanted oxygen on the electrical and optical properties of amorphous germanium", International Conference on Radiation Effects in Semiconductors, 1976, p. 330-1, EEA81-1049.

3. Group III-V Semiconductors
A. Gallium Arsenide

961. Agashe, V. V., Gupta, S. C. and Jain, B. P., "Ion implantation into GaAs", Indian J. Pure & Appl. Phys., 17 (5), 287-9 (1979), EEA83-14284.
962. Alexandre, F., "Study of the surface segregation of alkali metal impurities implanted in gallium arsenide", J. Phys., 39 (6), 701-10 (1978), French, EEA81-33124.
963. Ambridge, T., "Carrier concentration profiling in ion-implanted gallium arsenide by using a modified electrochemical technique", Report BPO-RDR-652, 1977, 12 pp., CA91-31616.
964. Anderson, C. L. and Dunlap, H. L., "Low-temperature annealing behavior of GaAs implanted with Be", Appl. Phys. Lett., 35 (2), 178-80 (1979), EEA82-46926.
965. Anderson, C. L., Vaidyanathan, K. V., Dunlap, H. L. and Kamath, G. S., "Capless annealing of ion-implanted gallium arsenide by a melt-controlled ambient technique", J. Electrochem. Soc., 127 (4), 925-7 (1980), EEA83-34803.
966. Anderson, W. J. and Park, Y. S., "Electroreflectance measurements of lattice damage in ion implanted gallium arsenide", J. Appl. Phys., 47 (7), 3094-8 (1976), CA85-135755.
967. Anderson, W. J. and Park, Y. S., "Flux and fluence dependence of implantation disorder in GaAs substrates", J. Appl. Phys., 49 (8), 4568-70 (1978), EEA81-45932.
968. Antell, G. R., "The effect of the anneal ambient on implanted GaAs and the occurrence of compensated regions in Si implants", Appl. Phys. Lett., 30 (8), 432-4 (1977), EEA80-24863.

969. Aoki, K., Gamo, K., Masuda, K. and Namba, S., "Photoluminescence study of cadmium-ion implanted n-gallium arsenide", Jpn. J. Appl. Phys., 15 (1), 145-9 (1976), CA84-67378.
970. Aoki, K., Gamo, K., Masuda, K. and Namba, S., "Depth distribution of defects in Mg-ion and Cd-ion implanted GaAs", Jap. J. Appl. Phys., 15 (2), 405-6 (1976), PA79-31708.
971. Asbeck, P. M., Tandon, J. and Welch, B. M., "Effects of Cr redistribution on electrical characteristics of ion-implanted semi-insulating GaAs", IEEE Electron Device Lett., EDL-1 (3), 35-7 (1980), EEA83-29135.
972. Barnes, P. A., Leamy, H. J., Poate, J. M., Ferris, S. D., Williams, J. S. and Celler, G. K., "Ohmic contacts produced by laser-annealing Te-implanted GaAs", AIP Conf. Proc., no. 50, 647-52 (1978), EEA82-38125.
973. Badawi, M. H., Sealy, B. J., Stephens, K. G., and Akintunde, J. A., "Improvement in electrical properties of laser annealed ion implanted GaAs", Jpn. J. Appl. Phys., 19, suppl. 19-1, 139-44 (1979), EEA83-48703.
974. Beneking, H., "Ion implantation for semiconductors", Patent German 2,733,146, Publ. January 1979, CA90-113976.
975. Bishop, S. G., Comas, J., Sundaram, S. and McCombe, B. D., "Effects of implantant depth distribution on photolaminescence spectra in Be-implanted GaAs", Appl. Phys. Lett., 31 (12), 845-7 (1977), EEA81-14642.
976. Boissy, M. C. and Diguet, D., "Zinc ion implantation as a predeposition process in gallium arsenide", J. Electrochem. Soc., 125 (9), 1505-9 (1978), EEA82-12568.
977. Bozler, C. O., Donnelly, J. P., Lindley, W. T. and Reynolds, R. A., "Impurity gettering in semi-insulating gallium arsenide using ion-implantation damages", Appl. Phys. Lett., 29 (11), 698-9 (1976), EEA80-6561.
978. Brawn, J. R. and Grant, W. A., "Lattice location studies of GaAs implanted with Te, Tl, and Zr", Applications of Ion Beams to Materials, 1975, p. 59-63, Publ. 1976, PA79-69557.
979. Breault, S. D., "Electrical effects of three different encapsulants on ion implanted GaAs", Report AD-A064362/7SL, Air Force Inst. of Tech., Wright-Patterson AFB, Oh., 1978, 63 pp.
980. Butcher, D. N. and Sealy, B. J., "Changes in the thermal oxidation of gallium arsenide induced by ion implantation", Electron Lett., 15 (2), 51-2 (1979), EEA82-17410.
981. Butcher, D. N. and Sealy, B. J., "The effects of ion implantation on the thermal oxidation of GaAs", Radiat. Eff., 48 (1-4), 203-6 (1980), EEA83-53683.
982. Campisano, S. U., Foti, G., Rimini, E., Catalano, I., Eisen, F. and Nicolet, M. -A., "Laser reordering of implanted amorphous layers in GaAs", Solid-State Electron, 21 (2), 485-8 (1978), EEA81-23877.

983. Campisano, S. U., Foti, G., Rimini, E., Eisen, F. H., Tseng, W. F., Nicolet, M. -A. and Tandon, J. L., "Laser pulse annealing of ion-implanted GaAs", J. Appl. Phys., 51 (1), 295-8 (1980), EEA83-23647.
984. Chang, T. and Teng, H., "The fabrication of an n-GaAs thin layer by means of sulphur-ion implantation", Applications of Ion Beams to Materials, 1975, p. 96-103, Publ. 1976, PA79-69562.
985. Chelnokov, V. E., Smirnova, N. S. and Suvorov, A. V., "Study of mixed crystals on the basis of GaAs produced by ion beam synthesis", Radiat. Eff., 48 (1-4), 1-6 (1980), EEA83-53654.
986. Comas, J. and Plew, L., "Beryllium and sulfur ion-implanted profiles in GaAs", J. Electron. Mater., 5 (2), 209-21 (1976), EEA79-46046.
987. Davies, D. E., "Implantation of GaAs and other III-V compounds", Proceedings of the 4th Conference on the Scientific and Industrial Applications of Small Accelerators, 1976, p. 420-9, EEA80-36260.
988. Davies, D. E., Kennedy, J. K. and Ludington, C. E., "Comparison of group IV and VI doping by implantation in GaAs", J. Electrochem. Soc., 122 (19), 1374-7 (1975), PA79-2736.
989. Dietrich, H. B., "Compensation in gallium arsenide by ion bombardment", Rep. NRL Prog., 3-6 (1976), CA85-39695.
990. Dmitruk, N. L., Zuev, V. A. and Litovchenko, V. G., Electrophysical and luminescent properties of implanted GaAs", Proceedings of the International Conference on Ion Beam Modification of Materials, 1978, p. 1111-20.
991. Donnelly, J. P., "Ion implantation in GaAs", Proceedings of the 6th International Symposium on Gallium Arsenide and related compounds, Pt. II, 1976, p. 166-90, EEA81-5925.
992. Donnelly, J. P., Bozler, C. O. and Lindley, W. T., "Low-dose n-type ion implantation into Cr-doped GaAs substrates", Solid-State Electron., 20 (3), 273-6 (1977), EEA80-17177.
993. Donnelly, J. P., Leonberger, F. J. and Bozler, C. O., "Uniform-carrier-concentration p-type layers in GaAs produced by beryllium ion implantation", Appl. Phys. Lett., 28 (12), 706-8 (1976), PA79-69513.
994. Donnelly, J. P., Lindley, W. T. and Hurwitz, C. E., "Reproducible, high carrier concentration implantations of selenium and silicon into GaAs", IEEE Trans. Electron Devices, ED-22 (11), 1065 (1975), EEA79-4547.
995. Dumoulin, J. D., "Depth-resolved cathodoluminescence on the effects of cadmium implantation and annealing in gallium arsenide", Report AD-A034029, Air Force Inst. Technol., Wright-Patterson AFB, Ohio, 1976, 75 pp., CA86-198788.
996. Duncan, W. M. and Schreiner, A. F., "Surface state study of ion implanted GaAs (Se) from photoreflectance", Solid State Commun., 31 (6), 457-9 (1979), PA82-91465.
997. Eisen, F. H., "Recent developments in ion-implantation doping of GaAs", J. Vac. Sci. & Technol., 14 (1), 530 (1977), EEA80-28826.

998. Eisen, F. H., Higgins, J. A., Immorlica, A. A., Kuvas, R. L. and Ludington, B. W., "Investigation of technological problems in GaAs", Report AD-A030599/5SL, Rockwell International, Thousand Oaks, Calif. Science Center, 1976, 141 pp.

999. Eisen, F. H., Welch, B., Gamo, K., Inada, T., Mueller, H., Nicolet, M. A. and Mayer, J. W., "Sulphur, selenium, and tellurium implantation in GaAs", Applications of Ion Beams to Materials, 1975, p. 64-8, Publ. 1976, PA79-69558.

1000. Eisen, F. H., Welch, B. M., Mueller, H., Gamo, K., Inada, T. and Mayer, J. W., "Tellurium implantation in GaAs", Solid-State Electron., 20 (3), 219-23 (1977), EEA80-17176.

1001. Elliott, C. R., Ambridge, T. and Heckingbottom, R., "Ion implantation damage in GaAs: a TEM study of the variation with ion species and stoichiometry", Solid-State Electron, 21 (6), 859-63 (1978), EEA81-33275.

1002. Emerson, N. G. and Sealy, B. J., "Laser annealing of low dose Se-implanted GaAs studied by d.l.t.s.", Electron. Lett., 16 (13), 512-14 (1980), EEA83-39742.

1003. Fan, J. C. C., Donnelly, J. P., Bozler, C. O. and Chapman, R. L., "Annealing of Se-implanted GaAs and InP by scanned Nd:YAG laser irradiation", Gallium Arsenide and Related Compounds 1978, p. 472-83, EEA83-9712.

1004. Favennec, P. N., "Semi-insulating layers of GaAs by oxygen implantation", J. Appl. Phys., 47 (6), 2532-6 (1976), PA79-73301.

1005. Favennec, P. N., Henry, L. and L'Haridon, H., "On the ion implantation of the group VI impurities into GaAs", Solid-State Electron., 21 (5), 705-10 (1978), EEA81-33274.

1006. Gamo, K., Inada, T., Krekeler, S., Mayer, J. W., Eisen, F. H. and Welch, B. M., "Selenium implantation in GaAs", Solid-State Electron., 20 (3), 213-17 (1977), EEA80-24880.

1007. Gamo, K., Katano, F., Yuba, Y., Murakami, K. and Namba, S., "Laser-annealing of Te ion implanted GaAs", AIP Conf. Proc., no. 50, 591-6 (1978), EEA82-38234.

1008. Gavrilov, A. A., Zelevinskaya, V. M., Kachurin, G. A., Pridachin, N. B. and Smirnov, L. S., "Orientational effects in the implantation of Te ions in gallium arsenide at elevated temperatures", Sov. Phys.-Semicond., 9 (11), 1435-6 (1975), PA79-66139.

1009. Gecim, H. S., Sealy, B. J. and Stephens, K. G., "Carrier removal after H_1^+, H_2^+ or H_3^+ implants into GaAs", Electron. Lett., 12 (25), 668-9 (1976), EEA80-6490.

1010. Gecim, H. S., Sealy, B. J. and Stephens, K. G., "Carrier removal profiles from oxygen implanted GaAs", Electron. Lett., 14 (10), 306-8 (1978), EEA81-33120.

1011. Gecim, H. S., Sealy, B. J., Stephens, K. G. and Ono, Y., "Carrier removal by implanted ions in GaAs", Proceedings of the International Conference on Ion Beam Modification of Materials, 1978, p. 349-56.

1012. Gibbons, J. F., "The use of ion implantation for the control of impurity diffusion in (group) III-V compound semiconductors", Report AD-A018760, Stanford Electron. Lab., Stanford Univ., Stanford, Calif., 1975, 16 pp., CA84-158767.
1013. Giorgianni, U., Grasso, V., Mondio, G., Perillo, P. and Saitta, G., "Optical transistions in the modulated reflectivity spectrum of ion-implanted GaAs", Can. J. Phys., 57 (6), 917-19 (1979), PA82-73972.
1014. Golovchenko, J. A. and Venkatesan, T. N. C., "Annealing of Te-implanted GaAs by ruby laser irradiation", Appl. Phys. Lett., 32 (3), 147-9 (1978), EEA81-19286.
1015. Grasso, V., Mondio, G., Saitta, G., Campisano, S. U., Foti, G. and Rimini, E., "Optical reflectivity of ion-implanted amorphous GaAs", Appl. Phys. Lett., 33 (7), 632-4 (1978), EEA82-4189.
1016. Greene, P. D., "Transmutation doping of GaAs by thermal neutrons", Solid State Commun., 32 (4), 325-6 (1979), EEA83-4433.
1017. Grote, N., Krautle, H. and Beneking, H., "Oxygen implantation in GaAs substrates: effect on LPE growth and electrical properties", Gallium Arsenide and Related Compounds 1978, p. 24-7, EEA83-9713.
1018. Hara, T. and Inada, T., "Trends in ion implantation in gallium arsenide". Solid State Technol., 22 (11), 69-74 (1979), EEA83-19005.
1019. Harris, T. J., Sealy, B. J. and Surridge, R. K., "Effect of channeling on the electrical properties of donor implanted GaAs", Electron Lett., 12 (25), 664-5 (1976), EEA80-3456.
1020. Harrison, H. B. and Martin, A. L., "Electrical properties of proton and deuterium ion implanted n-type GaAs", J. Appl. Phys., 51 (5), 2935-6 (1980), EEA83-43898.
1021. Hashimoto, M., Tamura, S., Yamagachi, J., Yokota, K. and Katayama, S., "A study of annealing effects for the lattice , expansion in ion-implanted GaAs", Technol. Rep. Kansai Univ., no. 17, 81-6 (1975), EEA79-32911.
1022. Houghton, A. J., "Switching conduction in ion-irradiated layers in GaAs", Electron. Lett., 14 (4), 102-3 (1978), EEA81-19290.
1023. Houghton, A. J. N., "Diffusion of zinc and silver on ion-implanted gallium arsenide", Thesis, Univ. Nottingham, England, 1979, EEA83-9709.
1024. Howes, M. J. and Morgan, D. V., "Influence of thin implanted layers on the properties of transferred electron devices", Applications of Ion Beams to Materials, 1975, p. 91-5, Publ. 1976, PA79-69561.
1025. Hubler, G. K., Comas, J. and Plew, L., "Profiles of ion-implanted Be in GaAs by means of (p,α) nuclear reaction and SIMS methods", Nucl. Instrum. & Methods, 149 (1-3), 635-8 (1978), EEA82-809.

1026. Hutchby, J. A. and Vaidyanathan, K. V., "Temperature dependence of electrical properties in Be-implanted semi-insulating GaAs", J. Appl. Phys., 48 (6), 2559-64 (1977), EEA80-36056.
1027. Immorlica, A. A. and Eisen, F. H., "Capless annealing of ion-implanted GaAs", Appl. Phys. Lett., 29 (2), 94-5 (1976), PA79-73294.
1028. Inada, T., Kato, S., Maeda, Y. and Tokunaga, K., "Doping profiles in Zn-implanted GaAs after laser annealing", J. Appl. Phys., 50 (9), 6000-2 (1979), EEA83-641.
1029. Inada, T., Kato, S., Ohkubo, T. and Hara, T., "Dual species ion implantation in GaAs", Radiat. Eff., 48 (1-4), 91-6 (1980), EEA83-53657.
1030. Inada, T., Miwa, H., Kato, S., Kobayashi, E., Hara, T. and Mihara, M., "Annealing of Se-implanted GaAs with an oxygen-free CVD Si_3N_4 encapsulant", J. Appl. Phys., 49 (8), 4571-3 (1978), EEA81-45949.
1031. Inada, T., Ohkubo, T., Kato, S., Kitahara, M., Kanda, Y. and Hura, T., "GaAs Hall-effect devices fabricated by ion-implantation technique", Electron Lett., 14 (16), 503-5 (1978).
1032. Inada, T., Tokunaga, K. and Taka, S., "Pulsed electron-beam annealing of selenium-implanted gallium arsenide", Appl. Phys. Lett., 35 (7), 546-8 (1979), EEA83-732.
1033. Itoh, T., Tsuchaya, T. and Takeuchi, M., "Carrier compensation of n-GaAs by oxygen ion implantation", Jap. J. Appl. Phys., 15 (11), 2277-8 (1976), EEA80-6582.
1034. Jervis, T. R., Woodard, D. W. and Eastman, L. F., "Effects of ion implantation on deep levels in GaAs", Electron. Lett., 15 (20), 619-21 (1979), EEA83-734.
1035. Kachare, A. H., Spitzer, W. G. and Fredrickson, J. E., "Refractive index of ion-implanted GaAs", J. Appl. Phys., 47 (9), 4209-12 (1976), PA80-1671.
1036. Kachare, A. H., Spitzer, W. G., Fredrickson, J. E. and Euler, F. K., "Measurements of layer thicknesses and refractive indices in high-energy ion-implanted GaAs and GaP", J. Appl. Phys., 47 (12), 5374-81 (1976), PA80-24837.
1037. Kasahara, J. and Arai, M., "Capless anneal of ion-implanted GaAs in controlled arsenic vapor", J. Appl. Phys., 50 (1), 541-3 (1979), EEA82-30376.
1038. Kasahara, J., Arai, M. and Watanabe, N., "Effect of arsenic partial pressure on capless anneal of ion-implanted GaAs", J. Electrochem. Soc., 126 (11), 1997-2001 (1979), EEA83-14293.
1039. Kawasaki, Y. and Nojima, S., "Be ion implantation into GaAs", Proceedings of the International Conference on Ion Beam Modification of Materials, 1978, p. 411-20.
1040. Kim, Q. and Park, Y. S., "Characterization of ion-implanted GaAs by ellipsometry", J. Appl. Phys., 51 (4), 2024-9 (1980), PA83--79819.

1041. Kim, Q., Park, Y. S., Mason, R. S., Luke, T. E., Hengehold, R. L. and Yeo, Y. K., "Laser annealing of ion-implanted GaAs", AIP Conf. Proc., no. 50, 597-602 (1978), EEA82-38235.
1042. Klein, P. B., Comas, J. and Bishop, S. G., "Donor-acceptor pair radiative recombination from ion-implanted donors and acceptors in GaAs", J. Appl. Phys., 50 (6), 4327-9 (1979), PA82-91489.
1043. Komarov, F. F. and Tashylkov, I. S., "Formation of defects in gallium arsenide on implantation of phosphorus ions at various temperatures", Sov. Phys.-Semicond., 11 (10), 1156-7 (1977), EEA81-33279.
1044. Komarov, F. F. and Tashlykov, I. S., "Peculiarities of distribution of defects and introduced impurity in P^+-implanted GaAs crystals", Proceedings of the International Conference on Ion Beam Modification of Materials, 1978, p. 1269-80.
1045. Kozeikin, B. V., Frolov, I. A. and Vysotskii, S. A., "Anomalous behaviour of impurities in ion-implantation-doped epitaxial n-type GaAs", Sov. Phys.-Semicond., 11 (5), 577-8 (1977), EEA81-10279.
1046. Kular, S. S. and Sealy, B. J., "Laser annealing of zinc implanted GaAs", Solid-State Electron, 23 (8), 875-80 (1980), EEA83-48596.
1047. Kular, S. S. and Sealy, B. J. and Stephens, K. G., "Electrical profiles from zinc implanted GaAs", Electron. Lett., 14 (1), 2-4 (1978), EEA81-14481.
1048. Kular, S. S., Sealy, B. J. and Stephens, K. G., "Comparison of electrical profiles from hot and cold implantations of zinc ions into GaAs", Electron. Lett., 14 (1), 22-3 (1978), EEA81-14482.
1049. Kular, S. S., Sealy, B. J., Stephens, K. G., Chick, D. R., Davis, Q. V. and Edwards, J., "Pulsed laser annealing of zinc implanted GaAs", Electron. Lett., 14 (4), 85-7 (1978), EEA81-19289.
1050. Kuznetsov, O. N., Lezheiko, L. V., Lyubopytova, E. V. and Safronov, L. N., "Synthesis of solid solutions by implantation of Al^+ and P^+ ions into GaAs", Sov. Phys.-Semicond., 11 (8), 851-3 (1977), EEA81-23750.
1051. Lee, D. H. and Malbon, R. M., "Ion-implanted silicon profiles in GaAs", Appl. Phys. Lett., 30 (7), 327-9 (1977), EEA80-24862.
1052. Lee, K. S., Ess, J. M., Littlejohn, M. A., Benson, R. B., Jr. and Comas, J., "A comparison between atomic concentration profiles and defect density profiles in GaAs annealed after implantation with beryllium", J. Electron. Mater., 9 (1), 185-96 (1980), EEA83-18997.
1053. Lidow, A., Gibbons, J. F., Deline, V. R. and Evans, C. A., Jr., "Ion-implanted selenium profiles in GaAs as measured by secondary ion mass spectrometry", Appl. Phys. Lett., 32 (1), 15-17 (1978), EEA81-19415.

1054. Lidow, A., Gibbons, J. F., Deline, V. R. and Evans, C. A., Jr., "Fast diffusion of elevated-temperature ion-implanted Se in GaAs as measured by secondary ion mass spectrometry", Appl. Phys. Lett., 32 (3), 149-51 (1978), EEA81-19418.
1055. Lidow, A., Gibbons, J. F., Deline, V. R. and Evans, C. A., Jr. "Ion implanted Se in GaAs", J. Appl. Phys., 51 (8), 4130-8 (1980), PA84-2469.
1056. Lidow, A., Gibbons, J. F. and Magee, T., "A double-layered encapsulant for annealing ion-implanted GaAs up to 1100°C", Appl. Phys. Lett., 31 (3), 158-61 (1977), EEA80-41243.
1057. Lin, M. S., Gamo, K., Masuda, K. and Namba, S., "Photoluminescence and electrical measurements on manganese ion-implanted gallium arsenide", Jpn. J. Appl. Phys., 15 (1), 53-6 (1976), CA84-68568.
1058. Liu, R.-Z. and Feng, W.-L., "LCAO-MO treatment for deep levels produced by ion implantation of oxygen in gallium arsenide", Int. J. Quantum Chem., 18 (2), 601-5 (1980), PA84-7552.
1059. Liu, S. G., Douglas, E. C., Wu, C. P., Magee, C. W., Narayan, S. Y., Jolly, S. T. and Kolondra, F., "Ion implantation of sulfur and silicon in GaAs", RCA Rev., 41 (2), 227-62 (1980), EEA83-53653.
1060. Liu, S. G., Wu, C. P. and Magee, C. W., "Annealing of ion-implanted GaAs with Nd:glass laser", AIP Conf. Proc., no. 50, 603-9 (1978), EEA82-38236.
1061. Malbon, R. M., Lee, D. H. and Whelan, J. M., "Annealing of ion-implanted GaAs in a controlled atmosphere", J. Electrochem. Soc., 123 (9), 1413-15 (1976), EEA79-46033.
1062. Mandal, R. P. and Scoble, W. R., "Ion-implanted gallium arsenide with proximity cap annealing", Gallium Arsenide and Related Compounds Symposium, 1978, p. 462-71, EEA83-9711.
1063. Martin, G. M., Berth, M. and Venger, C., "Role of substrate in electrical properties of GaAs implanted layers", Electron Lett., 16 (8), 278-9 (1980), EEA83-28952.
1064. Mason, R. S., "Laser annealing of ion implanted gallium arsenide", Report AD-A063961-7SL, Air Force Inst. of Tech., Wright-Patterson AFB, OH, 1978, 68 pp.
1065. Massics, J., Devoldere, P. and Linh, N. T., "Silver contact on GaAs (001) and InP (001)", J. Vac. Sci. & Technol., 15 (4), 1353-7 (1978), EEA82-733.
1066. Masuyama, A., Nicolet, M.-A., Golecki, I., Tandon, J. L., Sanada, D. K. and Washburn, J., "Steady-state thermally annealed GaAs with room-temperature-implanted Si", Appl. Phys. Lett., 36 (9), 749-51 (1980), EEA83-34629.
1067. Matsumura, H. and Stephens, K. G., "Electrical measurement of the lateral spread of the proton isolation layer in GaAs", J. Appl. Phys., 48 (7), 2779-83 (1977), EEA80-41387.
1068. Mayer, J., Tseng, W. and Eisen, F., "Recrystallization of implanted amorphous layers in GaAs", J. Vac. Sci. & Technol., 14 (4), 1006 (1977), EEA81-1197.

1069. McLevige, W. V., Helix, M. J., Vaidyanathan, K. V. and Streetman, B. G., "Electrical profiling and optical activation studies of Be-implanted GaAs", J. Appl. Phys., 48 (8), 3342-6 (1977), EEA81-1058.
1070. Min-Shyong, L., "Amorphous layer formation in GaAs by ion implantation", Electron. Lett., 14 (21), 695-7 (1978), EEA82-692.
1071. Molnar, B., "Close-contact annealing of ion-implanted GaAs and InP", Appl. Phys. Lett., 36 (11), 927-30 (1980).
1072. Molnar, B. and Kennedy, T. A., "Evaluation of S- and Se-implanted GaAs by contactless mobility measurement", J. Electrochem. Soc., 125 (8), 1318-20 (1978), EEA82-802.
1073. Monemar, B. and Blum, J. M., "Optical characterization of deep O implants in GaAs", J. Appl. Phys., 48 (4), 1529-37 (1977), EEA80-24869.
1074. Moore, J. A., Carter, G. and Tinsley, A. W., "A comparison of the radiation damage produced in gallium arsenide by monoatomic and diatomic arsenic implants", Radiat. Eff., 25 (1), 49-51 (1975), PA79-5475.
1075. Mueller, J. E., Kellner, W., Kniepkamp, H., Haas, E. W. and Fischer, G., "Doping of semi-insulating and n-type GaAs by neutron transmutation", J. Appl. Phys., 51 (6), 3178-80 (1980).
1076. Mullins, B. W., "Characterization of laser annealing of ion implanted GaAs and Si using optical reflectivity", Report AD-A080243-9, Air Force Inst. of Tech., Wright-Patterson AFB, OH, 1979, 66 pp.
1077. Namba, S., Kawabe, M., Kanzaki, N. and Masuda, K., "Photoluminescence measurement of ion-etched GaAs surface", J. Vac. Sci. Technol., 12 (6), 1348-51 (1975).
1078. Narayanan, G. H. and Spitzer, W. G., "The structural characteristics of radiation damage produced by high energy (2.7 MeV) ion implantation in GaAs", J. Mater. Sci., 13 (11), 2418-28 (1978), EEA82-4303.
1079. Nojima, S., "Physical property of disordered-GaAs produced by ion implantation", Jpn. J. Appl. Phys., 18 (6), 1079-86 (1979), EEA82-38067.
1080. Nojima, S. and Kawasaki, Y., "Property of disordered GaAs by ion implantation", Jpn. J. Appl. Phys., 17 (8), 1455-6 (1978), EEA81-45838.
1081. Nojima, S. and Kawasaki, Y., "Annealing characteristics of Be ion implanted GaAs", Jpn. J. Appl. Phys., 17 (10), 1845-50 (1978), EEA82-4304.
1082. Nozaki, T., "Si ion implantation into GaAs", Jap. J. Appl. Phys., 15 (10), 1951-9 (1976), EEA80-618.
1083. Park, Y. S., Grant, J. T. and Haas, T. W., "The determination of sulfur-ion-implantation profiles in GaAs using Auger electron spectroscopy", J. Appl. Phys., 50 (2), 809-12 (1979), EEA82-30379.

1084. Park, Y. S., Theis, W. M. and Grant, J. T., "Characterization of ion implants in GaAs by AES and GDOS", Appl. Surf. Sci., 4 (3-4), 445-55 (1980), EEA83-28949.
1085. Pianetta, P. A., Stolte, C. A. and Hansen, J. L., "Nonalloyed ohmic contacts to electron-beam-annealed Se-ion-implanted GaAs", Appl. Phys. Lett., 36 (7), 597-9 (1980), EEA83-29056.
1086. Plew, L. and Comas, J., "Effect of annealing on Be ion-implanted profiles in GaAs", Electrochemical Society Spring Meeting (Extended Abstracts), 1976, p. 352, PA79-87927.
1087. Price, P. B., Gadeken, L. L., Mak, H. B. and Ewan, G. T., "X-ray and Rutherford backscattering yields from channeled helium ions in gallium arsenide", Radiat. Eff., 39 (3-4), 205-12 (1978), CA90-160368.
1088. Puangthum, S., "Glow discharge optical spectroscopy of ion implanted gallium arsenide", Report AD-A080413-8, Air Force Inst. of Tech., Wright-Patterson AFB, OH, 1979, 86 pp.
1089. Quiesup, K. and Park, Y. S., "Characterization of ion-implanted GaAs by ellipsometry", J. Appl. Phys., 51 (4), 2024-29 (1980).
1090. Rao, E. V. K., "Optical properties of oxygen-implanted GaAs", Phys. Status Solidi A, 33 (2), 683-90 (1976).
1091. Rao, E. V. K., Duhamel, N., Favennec, P. N. and L'Haridon, H., "Investigation of compensation in implanted n-GaAs", J. Appl. Phys., 49 (7), 3898-905 (1978), EEA81-42346.
1092. Rosinski, W. and Komicz, T., "Properties of Mn-ion implanted p-n junctions in n-type GaAs", Electron Technol., 7 (3-4), 69-72 (1974), PA79-22419.
1093. Sadana, D. K. and Booker, G. R., "TEM structural studies on Se^+ implanted GaAs", Proceedings of the International Conference on Ion Beam Modification of Materials, 1978, p. 1131-42.
1094. Sadana, D. K., Booker, G. R., Sealy, B. J., Stephens, K. G. and Badawi, M. H., "Correlation between structural and electrical profiles in ion-implanted GaAs", Proceedings of the International Conference on Ion Beam Modification of Materials, 1978, p. 479-86.
1095. Sakurai, T., Nanbu, K. and Furuya, T., "Sulfur ion-implanted GaAs", Fujitsu Sci. & Tech. J., 12 (3), 121-30 (1976), EEA80-6575.
1096. Sealy, B. J., Badawi, M. H., Kular, S. S. and Stephens, K. G., "Laser annealing of ion implanted GaAs", AIP Conf. Proc., no. 50, 610-15 (1978), EEA82-38237.
1097. Sealy, B. J., Bell, E. C., Surridge, R. K. and Stephens, K. G., "Dual-ion implantation into GaAs", Applications of Ion Beams to Materials, 1975, p. 75-80, Publ. 1976, PA79-69560.
1098. Sealy, B. J., Kular, S. S. and Stephens, K. G., "Electrical properties of laser-annealed donor-implanted GaAs", Electron. Lett., 14 (22), 720-1 (1978), EEA82-4191.
1099. Sealy, B. J., Kular, S. S., Stephens, K. G., Sadana, D. and Booker, G. R., "Laser annealing of ion implanted GaAs", Radiat. Eff., 48 (1-4), 121-4 (1980), EEA83-53659.

1100. Sealy, B. J. and Surridge, R. K., "Thermal pulse annealing of donor implanted GaAs", Proceedings of the International Conference on the Ion Beam Modification of Materials, 1978, p. 487-96.
1101. Sealy, B. J., Surridge, R. K., Bell, E. C. and D'Cruz, A. D. E., "Donor activity in ion-implanted GaAs", Applications of Ion Beams to Materials, 1975, p. 69-74, Publ. 1976, PA79-69559.
1102. Sealy, B. J., Surridge, R. K., Kular, S. S. and Stephens, K. G., "Pulse annealing of ion-implanted GaAs", International Conference on Defects and Radiation Effects in Semiconductors, 1978, p. 476-81, EEA83-14309.
1103. Shah, N. J., Ahmed, H., Sanders, I. R. and Singleton, J. F., "Activation of low dose silicon implants in GaAs by multiply scanned electron beams", Electron. Lett., 16 (11), 433-6 (1980), EEA83-34793.
1104. Shin, B. K., "Carbon-ion-implanted gallium arsenide", Appl. Phys. Lett., 29 (7), 438-40 (1976), EEA80-13450.
1105. Shin, B. K., Ehret, J. E., Park, Y. S. and Stefiniw, M., "Dual implantation of C^+ and Ga^+ ions into GaAs", J. Appl. Phys., 49 (5), 2988-90 (1978), EEA81-38122.
1106. Shin, B. K., Look, D. C., Park, Y. S. and Ehret, J., "Hall-effect measurements in Cd-implanted GaAs", J. Appl. Phys., 47 (4), 1574-9 (1976), EEA79-28350.
1107. Shin, B. K. and Park, Y. S., "An abrupt dopant profile in GaAs produced by Te implantation", J. Electrochem. Soc., 123 (10), 1588-9 (1976), EEA80-617.
1108. Shinoda, D., "Compound semiconductor layer of high resistivity", Patent USA 3936321, Publ. February 1976, Japanese, CA84-172867.
1109. Siu, D. P. and Immorlica, A. A., Jr., "Comparison of doping profiles in capless annealed and dielectrically capped-annealed ion implanted GaAs", J. Electron. Mater., 9 (5), 857-68 (1980), EEA83-53652.
1110. Solt, R. M., "Damage studies of ion-implanted GaAs using conductivity profiles", Report AD-A064067-2SL, Air Force Inst. of Tech., Wright-Patterson AFB, OH, 1978, 67 pp.
1111. Stephens, K. G. and Sealy, B. J., "Use of ion implantation in future GaAs technology", Microelectronics, 9 (2), 13-18 (1978), EEA82-220287.
1112. Stolte, C. A., "Device quality n-type layers produced by ion implantation of Te and S into GaAs", 1975 International Electron Devices Meeting, Technical Digest, p. 585-7, EEA80-3469.
1113. Stoneham, E. B., Patterson, G. A. and Gladstone, J. M., "Formation of heavily doped n-type layers in GaAs by multiple ion implantation", J. Electron. Mater., 9 (2), 371-83 (1980), EEA83-29138.

1114. Surridge, R. K. and Sealy, B. J., "Active layers for device applications by using high-energy selenium implantation into GaAs", Electron. Lett., 13 (8), 233-4 (1977), EEA80-17169.
1115. Surridge, R. K. and Sealy, B. J., "A comparison of Sn-, Ge-, Se and Te-ion-implanted GaAs", J. Phys. D, 10 (6), 911-17 (1977), EEA80-17108.
1116. Surridge, R. K., Sealy, B. J., D'Cruz, A. D. E. and Stephens, K. G., "Annealing kinetics of donor ions implanted into GaAs", Proceedings of the 6th International Symposium on Gallium Arsenide and Related Compounds, Pt. I, 1976, p. 161-7, EEA81-5922.
1117. Takai, M., Gamo, K., Masuda, K. and Namba, S., "Lattice site location of cadmium and tellurium implanted in gallium arsenide", Jap. J. Appl. Phys., 14 (12), 1935-42 (1975), PA79-18896.
1118. Takai, M., Gamo, K., Yagita, H., Masuda, K., Namba, S. and Mizobuchi, A., "Use of ion-induced X-rays to locate ion-implanted impurities in gallium arsenide", Nucl. Instrum. & Methods, 149 (1-3), 457-9 (1978), EEA82-807.
1119. Tandon, J. L. and Eisen, F. H., "Pulsed annealing of implanted semi-insulating GaAs", AIP Conf. Proc., no. 50, 616-22 (1978), EEA82-38238.
1120. Tandon, J. L., Golecki, I., Nicolet, M.-A., Sadana, D. K. and Washburn, J., "Pulsed electron beam induced recrystallization and damage in GaAs", Appl. Phys. Lett., 35 (11), 867-9 (1979), EEA83-14279.
1121. Tandon, J. L., Nicolet, M.-A. and Eisen, F. H., "Silicon implantation in GaAs", Appl. Phys. Lett., 34 (2), 165-7 (1979), EEA82-26744.
1122. Tandon, J. L., Nicolet, M.-A., Tseng, W. F., Eisen, F. H., Campisano, S. U., Foti, G. and Rimini, E., "Pulsed-laser annealing of implanted layers in GaAs", Appl. Phys. Lett., 34 (9), 597-9 (1979), EEA82-34264.
1123. Tosic, M. M., "The calculation of range distribution of implanted dopants in GaAs and GaP", Fizika, 10, suppl. no. 2, 93-7 (1978), EEA82-30371.
1124. Toulouse, B. and Favennec, P. N., "Deep levels in boron implanted GaAs", Rev. Phys. Appl., 15 (4), 869-74 (1980), French, EEA83-28973.
1125. Toulouse, B., Favennec, P. N., Guivarc'h, A. and Pelous, G., "Physiochemical and resistivity studies of boron-implanted GaAs", Gallium Arsenide and Related Compounds Symposium, 1978, p. 501-9, EEA83-9545.
1126. Toulouse, B., Gauneau, M., Guivarc'h, A. and Pelous, G., "Channeling and secondary ion micro-analysis studies of 0.2-2 MeV ^{11}B implanted in GaAs", Proceedings of the International Conference on Ion Beam Modification of Materials, 1978, p. 525-36.

1127. Tremain, R. E., Jr., "Doping gallium arsenide by ion implantation and proton-enhanced diffusion", Thesis, Stanford University, 1976, 107 pp., Order No. 76-18807.
1128. Tseng, W., Dietrich, H., Davey, J., Christou, A. and Anderson, W. T., Jr., "The effects of ion implantation and pulsed electron beam anneal on Ge films grown epitaxially on <100> GaAs", J. Electron. Mater., 9 (3), 685-92 (1980), EEA83-34811.
1129. Tsu, R., Baglin, J. E., Lasher, G. J. and Tsang, J. C., "Laser induced damage and recrystallization of ion implanted GaAs by frequency-doubled Nd:YAG laser", AIP Conf. Proc., no. 50, 623-8 (1978), EEA82-38239.
1130. Tsuji, T. and Hasegawa, F., "Degradation of the doping profile of epitaxial GaAs layers due to an ion implantation process", IEEE Electron Device Lett., EDL-1 (6), 112-14 (1980), EEA83-39754.
1131. Tunkasiri, T. and Lewis, D., "Twinning in annealed tellurium-implanted gallium arsenide crystals", Radiat. Eff., 27 (1-2), 109-10 (1975), PA79-35463.
1132. Tunkasiri, T. and Lewis, D., "Surface effects due to encapsulation of ion-implanted GaAs", Radiat. Eff., 30 (3), 185-6 (1976), EEA80-623.
1133. Vasudev, P. K., Wilson, R. G. and Evans, C. A., Jr., "Chromium redistribution during thermal annealing of semi-insulating GaAs as a function of encapsulant and implant fluence", Appl. Phys. Lett., 36 (10), 837-40 (1980), EEA83-39739.
1134. Vasudev, P. K., Wilson, R. G. and Evans, C. A., "Damage gettering of Cr during the annealing of Cr and S implants in semi-insulating GaAs", Appl. Phys. Lett., 37 (3), 308-10 (1980), EEA83-53643.
1135. Venkatesan, T. N. C., Auston, D. H., Golovchenko, J. A. and Surko, C. M., "Study of surface crystallinity and stoichiometry of laser annealed GaAs using time resolved reflectivity and channeling", AIP Conf. Proc., no. 50, 629-34 (1978), EEA82-38240.
1136. Voltsit, V. V., Drazhan, A. V., Zuev, V. A., Ivaniichuk, M. T., Korbutyak, D. V. and Litovchenko, V. G., "Photoluminescence of gallium arsenide doped by silicon implantation", Sov. Phys. Semicond., 12 (10), 1211-13 (1978).
1137. Williams, J. S. and Austin,M. W., "Low-temperature epitaxial regrowth of ion-implanted amorphous GaAs", Appl. Phys. Lett., 36 (12), 994-7 (1980), PA83-92790.
1138. Williamson, K. R., "Glow discharge optical spectroscopy of ion implanted gallium arsenide", Report AD-A064419-5SL, Air Force Inst. of Tech., Wright-Patterson AFB, OH, 1978, 74 pp.
1139. Witmeyer, R. J., "A correlation of defect density and atomic concentration profiles of Be ion implanted layers in gallium-arsendide", Thesis, North Carolina State University at Raleigh, NC, 1978, 121 pp., Order No. 78-20070.

1140. Woodcock, J. M., "Laser pulse annealing of ion implanted gallium arsenide", Philips Res. Lab. (Redhill) Annu. Rev., 33-5 (1978), EEA83-9707.
1141. Woodcock, J. M. and Butler, J. M., "Highly doped implanted donor layers in laser annealed gallium arsenide", Radiat. Eff., 47 (1-4), 81-4 (1980), EEA83-44008.
1142. Woodcock, J. M. and Shannon, J. M., "Ion implantation into GaAs and related compounds", Acta Electron., 19 (2), 151-9 (1976), PA79-66110.
1143. Yokota, K., Tamura, S., Gamo, K., Namba, S., Masuda, K., Ishihara, S. and Kimura, I., "Ion-implanted arsenic profiles in GaAs encapsulated by SiO_2 and Si_3N_4", Jpn. J. Appl. Phys., 17 (10), 1881-2 (1978), EEA82-4197.
1144. Yokoyama, N., Ohikawa, S., Sakurai, T., Furuya, T. and Ishikawa, H., "Deep trap measurement on S^+-implanted layers in GaAs", Appl. Phys. Lett., 28 (7), 405-7 (1976), EEA79-24700.
1145. Yu, P. W. and Park, Y. S., "Temperature dependence of photoluminescence from Mg-implanted GaAs", J. Appl. Phys., 48 (6), 2434-41 (1977), EEA80-32382.
1146. Yuba, Y., Gamo, K. and Namba, S., "DLTS study of implanted GaAs", Proceedings of the International Conference on Ion Beam Modification of Materials, 1978, p. 557-66.
1147. Zelevinskaya, V. M., Kachurin, G. A. and Smirnov, L. S., "Interstitial diffusion of an impurity from ion-implanted layers", Sov. Phys.-Semicond., 9 (8), 1075-6 (1975), PA79-45089.
1148. Zelevinskaya, V. M., Kachurin, G. A., Safronov, L. N. and Smirnov, L. S., "Conversion of type of conduction stimulated by ion bombardment of silicon-doped GaAs", Sov. Phys.-Semicond., 10 (1), 99 (1976), EEA80-3402.
1149. Zuk, W. and Bazyluk, D., "The influence of radiation damage on the diffusion of P and Kr atoms implanted in GaAs monocrystals", International Conference on Radiation Effects in Semiconductors, 1976, p. 511-21, EEA81-5926.

B. Others

1150. Anderson, R. E., Wolford, D. J. and Streetman, B. G., "Nitrogen implantation in $GaAs_{1-x}P_x$. II. Annealing properties", J. Appl. Phys., 48 (6), 2453-62 (1977), EEA80-32385.
1151. Arnoldussen, T. C. and Greenstein, E., "Ion-implanted gallium-arsenide-phosphide surfaces", J. Electrochem. Soc., 124 (7), 1102-6 (1977), EEA80-36226.
1152. Bahir, G. and Kalish, R., "Dose rate and orientation dependence of damage induced by Xe and Cd implants in InSb", Radiat. Eff., 33 (4), 199-203 (1977), EEA81-1200.

1153. Belyi, I. M., Gumanskii, G. A., Karas, V. I., Lomako, V. M., Tashlykov, I. S., Tishkov, V. S. and Lenin, V. I., "Synthesis of $Al_xGa_{1-x}As$ and $GaAs_{1-x}P_x$ by implantation of Al^+ and P^+ ions into GaAs", Sov. Phys.-Semicond., 9 (10), 1326-7 (1975), PA79-58563.
1154. Blaut-Blachev, A. N., Gerasimenko, N. N., Lezheiko, L. V., Lyubopytova, E. V. and Obodnikov, V. I., "Nature of p-n conversion of ion-irradiated InSb crystals,", Sov. Phys.-Semicond., 14 (2), 306-10 (1980), EEA83-48714.
1155. Bogatyrev, V. A. and Kachurin, G. A., "Annealing of defects in ion-bombarded indium antimonide", Sov. Phys.-Semicond., 11 (7), 798-800 (1977), EEA81-14499.
1156. Bogatyrev, V. A., Kachurin, G. A. and Smirnov, L. S., "Implantation of ions in indium antimonide at elevated temperatures", Sov. Phys.-Semicond., 12 (1), 57-9 (1978), EEA82-8268.
1157. Bogatyrev, V. A., Kachurin, G. A. and Smirnov, L. S., "Diffusion of zinc out of implanted indium antimonide layers", Sov. Phys.-Semicond., 12 (5), 517-19 (1978), EEA82-8160.
1158. Bubulac, L. O., Barrowcliff, E. E., Tennant, W. E., Pasko, J. G., Williams, G., Andrews, A. M., Gheung, D. T. and Gertner, E. R., "Be ion implantation in InAsSb and GaInSb", Gallium Arsenide and Related Compounds Symposium, 1978, p. 519-29, EEA83-9714.
1159. Chatterjee, P. K., McLevige, W. V. and Streetman, B. G., "Electrical properties of Be-implanted $GaAs_{1-x}P_x$", Solid-State Electron., 19 (11), 961-4 (1976), EEA79-45987.
1160. Chatterjee, P. K. and Streetman, B. G., "Reduced lateral diffusion and reverse leakage in beryllium-implanted gallium arsenide phosphide ($GaAs_{1-x}P_x$) diodes", Solid-State Electron., 20 (4), 305-6 (1977), CA86-198798.
1161. Chernysheva, N. Y., Kachurin, G. A. and Bogatyrev, V. A., "Amorphisation and recrystallisation of InSb ion implanted layers", Phys. Status Solidi A, 47 (1), K5-8 (1978), PA81-63782.
1162. Chevrier, J., Huber, A. and Linh, N. T., "Zn-doped vapor-grown InP", J. Appl. Phys., 51 (1), 815-17 (1980), EEA83-23914.
1163. Churin, S. A. and Frolov, I. A., "Accumulation of radiation defects as a result of ion bombardment of GaP single crystals", Sov. Phys.-Semicond., 11 (2), 241-2 (1977), EEA80-36242.
1164. Cullis, A. G., Webber, H. C. and Robertson, D. S., "Laser annealing of ion-implanted indium phosphide", AIP Conf. Proc., no. 50, 653-8 (1978), EEA82-38241.
1165. Danilov, Y. A. and Tuluvchikov, V. S., "Anomalous radiation disordering of indium antimonide by ion implantation", Sov. Phys.-Semicond., 14 (1), 117-19 (1980), PA83-92553.
1166. Davies, D. E., Comer, J. J., Lorenzo, J. P. and Ryan, T. G., "Improved electrical mobilities from implanting InP at elevated temperatures", Appl. Phys. Lett., 35 (2), 192-4 (1979), EEA83-629.

1167. Davies, D. E., Kennedy, J. K. and Hawley, J. J., "Ion implantation in $InAs_xP_{1-x}$", IEEE Trans. Electron Devices, ED-22 (11), 1066 (1975), PA79-10483.
1168. Davies, D. E., Kennedy, J. K. and Hawley, J. J., "Electrical effects of implantation damage in $InAs_xP_{1-x}$", Applications of Ion Beams to Materials, 1975, p. 81-5, Publ. 1976, EEA79-32811.
1169. Davies, D. E. and Lorenzo, J. P., "Isolation and n-type conduction from implantation in InP", IEEE Trans. Electron Devices, ED-24 (9), 1205 (1977), EEA81-5916.
1170. Davies, D. E., Lorenzo, J. P. and Deane, M. L., "Compensation from implantation damage in InP", Appl. Phys. Lett., 31 (4), 256-8 (1977), EEA80-41378.
1171. Davies, D. E., Lorenzo, J. P. and Ryan, T. G., "n-type doping of indium phosphide by implantation", Solid-State Electron., 21 (7), 981-5 (1978), EEA81-49730.
1172. Davies, D. E., Lorenzo, J. P., Ryan, T. G. and Fitzgerald, J. J., "Pulse electron annealing of ion-implanted InP", Appl. Phys. Lett., 35 (8), 631-3 (1979), EEA83-9689.
1173. Davies, D. E., Potter, W. D. and Lorenzo, J. P., "Implantation and PH_3 ambient annealing of InP", J. Electrochem. Soc., 125 (11), 1845-8 (1978), EEA82-22083.
1174. Destefanis, G. L. and Gailliard, J. P., "Very efficient void formation in ion implanted InSb", Appl. Phys. Lett., 36 (1), 40-2 (1980), PA83-44237.
1175. Dobbs, B. C., Anderson, W. J. and Park, Y. S., "Investigation of ion-implanted GaP layers by ellipsometry", J. Appl. Phys., 48 (12), 5052-6 (1977), EEA81-23865.
1176. Donnelly, J. P. and Armiento, C. A., "Beryllium-ion implantation in InP and $In_{1-x}Ga_xAs_yP_{1-y}$", Appl. Phys. Lett., 34 (1), 96-9 (1979), EEA82-22078.
1177. Donnelly, J. P. and Hurwitz, C. E., "Proton bombardment in InP", Solid-State Electron., 20 (8), 727-30 (1977), EEA80-41274.
1178. Donnelly, J. P. and Hurwitz, C. E., "Ion implantation in InP", IEEE Trans. Electron Devices, ED-24 (9), 1204-5 (1977), EEA81-5915.
1179. Donnelly, J. P. and Hurwitz, C. E., "Ion-implanted n- and p-type layers in InP", Appl. Phys. Lett., 31 (7), 418-20 (1977), EEA81-5912.
1180. Donnelly, J. P. and Hurwitz, C. E., "High-resistivity layers in n-InP produced by Fe ion implantation", Solid-State Electron., 21 (2), 475-8 (1978), EEA81-28423.
1181. Eisen, F. H., "Ion implantation in III-V compounds", Radiat. Eff., 47 (1-4), 99-115 (1980), EEA83-44012.
1182. Favennec, P. N., Henry, L. and Janicki, T., "Implantation of selenium into $GaAl_{1-x}As_x$", Electron. Lett., 13 (12), 338-9 (1977), EEA80-32381.

1183. Flagmeyer, R., Ascheron, C., Geist, V., Ullrich, H.-J. and Stephan, D., "Determination of lattice expansion in proton-bombarded GaP single crystals,", Proceedings of the International Conference on Ion Beam Modification of Materials, 1978, p. 1627-38.

1184. Gamo, K., Takai, M., Yagita, H., Takada, N., Masuda, K., Namba, S. and Mizobuchi, A., "Implantation temperature for III-V compound semiconductors", J. Vac. Sci. & Technol., 15 (3), 1086-8 (1978), EEA81-42352.

1185. Gamo, K., Yagita, H., Takai, M. and Namba, S., "Ion implantation of Zn, Se and Cd in BP crystals", Radiat. Eff., 47 (1-4), 45-9 (1980), EEA83-44007.

1186. Gueseva, M. I., Mansurova, A. N., Tikhonov, V. G. and Khorvat, S. N., "Investigation of the Hall effect in indium antimonide doped by implantation of sulfur ions", Sov. Phys.-Semicond., 10 (8), 872-3 (1976), EEA80-13383.

1187. Holm, N. E., Nylandsted, L. A., Deutch, B. I., Weyer, G. and Albany, J. H., "Mossbauer spectroscopy of ion-implanted $^{119}Sn^{m}$ in GaAs", International Conference on Defects and Radiation Effects in Semiconductors, 1978, Publ. 1979, p. 573-8, PA83-29667.

1188. Inada, T. and Ohnuki, Y., "Anodic oxidation of GaP in N-methyl-acetamide for electrical profiling of ion-implanted GaP", J. Electrochem. Soc., 127 (1), 68-74 (1980), EEA83-23648.

1189. Kachare, A. H., Cherlow, J. M., Yang, T. T., Spitzer, W. G. and Euler, F. K., "Infrared reflection and Raman scattering of ion-implanted nitrogen in gallium phosphide", J. Appl. Phys., 47 (1), 161-73 (1976), EEA79-20298.

1190. Kasahara, J., Gibbons, J. F., Magee, T. J. and Peng, J., "Ion implantation of sulfur in Cr-doped InP at room temperature", J. Appl. Phys., 51 (8), 4119-24 (1980), PA84-2477.

1191. Korshunov, A. B., Kuznetsov, G. M., Makarov, A. G., Olenin, V. V. and Postnikov, I. V., "Investigation of isochronous annealing of p-type InSb irradiated with moderate-energy ions", Sov. Phys.-Semicond., 12 (5), 554-7 (1978), EEA82-12463.

1192. Lank, D. J., Dobbs, B. C. and Park, Y. S., "Electrical profiles of magnesium-ion-implanted GaP", J. Appl. Phys., 50 (3), pt. 1, 1318-24 (1979), EEA82-38062.

1193. Linden, M., Ejder, E. and Hellborg, R., "Channelling studies of as grown GaN",Phys. Status Solidi A, 45 (2), K167-70 (1978), EEA81-23746.

1194. Lorenzo, J. P., Davies, D. E. and Ryan, T. G., "Anodic oxidation and electrical carrier concentration profiles of ion-implanted indium phosphide", J. Electrochem. Soc., 126 (1), 118-21 (1979), CA90-79605.

1195. Makita, Y., Nojiri, H., Tsurushima, T., Tanoue, H. and Shimada, J., "Observation of in situ annealing in hot ion-implantation of nitrogen into $Al_xGa_{1-x}As$(x=0.53)", Appl. Phys. Lett. 35 (3), 293-5 (1979), EEA83-4276.
1196. McLevige, W. V., Vaidyanathan, K. V., Streetman, B. G., Comas, J. and Plew, L., "Annealing studies of Be-implanted $GaAs_{0.6}P_{0.4}$", J. Electron Mater, 7 (4), 547-58 (1978), EEA81-45839.
1197. Millea, M. F. and Silver, A. H., "Influence of the free surface on the electrical behaviour of metal contacts to p-InAs", J. Vac. Sci. & Technol., 15 (4), 1362-9 (1978), EEA82-735.
1198. Molnar, B., "Close-contact annealing of ion-implanted GaAs and InP", Appl. Phys. Lett., 36 (11), 927-9 (1980), EEA83-39741.
1199. Muller, G., Haubold, M., Schimko, R., Richter, C. -E. and Schwarz, G., "The diffusion of ion-implanted zinc in $GaAs_{1-x}P_x$", Phys. Status Solidi A, 42 (2), 579-83 (1977), EEA80-41395.
1200. Muller, G., Haubold, M., Schimko, R., Trapp, M. and Schwarz, G., "Investigations of the diffusion of implanted zinc in $GaAs_{1-x}P_x$ by ion microprobe", Phys. Status Solidi A, 49 (1), 279-84 (1978), EEA81-49728.
1201. Novak, I. I., Baptizmanskii, V. V., Smirnova, N. S. and Suvorov, A. V., "A Raman scattering investigation of $GaAs_{1-x}P_x$ and $Ga_{1-x}Al_xAs$ solid solutions synthesized by ion implantation", Sov. Phys.-Solid State, 20 (7), 1232-5 (1978), PA82-42972.
1202. Okabayashi, H., "Effects of electrically inactive ion implantation in $GaAs_{1-x}P_x$($x \simeq 0.38$) on photoluminescence intensity", Appl. Phys. Lett., 28 (9), 490-1 (1976).
1203. Petkov, M. D., "Zinc concentration in epitaxial layers of gallium phosphide doped with zinc and tellurium", Bulg. J. Phys., 6 (2), 178-83 (1979), EEA82-38242.
1204. Rothemund, W. and Fritzsche, C. R., "Radiation damage in ion-implanted GaP and $GaAs_{0.6}P_{0.4}$", Appl. Phys. Lett., 33 (5), 435-7 (1978), EEA81-49714.
1205. Rothemund, W. and Fritzsche, C. R., "Critical implantation temperature and annealing of indium phosphide", J. Vac. Sci. & Technol., 16 (3), 968-9 (1979), EEA82-47082.
1206. Shimada, T., Kato, Y., Shiraki, Y. and Komatsubara, K. F., "Amorphous GaP produced by ion implantation", J. Phys. & Chem. Solids, 37 (3), 305-13 (1976), PA79-27028.
1207. Shimada, T., Shiraki, Y. and Komatsubara, K. F., "Nitrogen centers in GaP produced by hot implantation", J. Phys. & Chem. Solids, 37 (3), 315-19 (1976), PA79-27029.
1208. Stoneham, E. B. and Gibbons, J. F., "Control of zinc diffusion in $GaAs_{0.6}P_{0.4}$ by ion implantation", IEEE Trans. Electron Devices, ED-22 (11), 1065-6 (1975), PA79-10482.
1209. Stoneham, E. B. and Gibbons, J. F., "Control of zinc diffusivity in $GaAs_{0.6}P_{0.4}$ by multiple implantation", J. Appl. Phys., 48 (12), 5086-91 (1977), EEA81-23866.

1210. Takai, M., Ishida, T., Masuda, K., Gamo, K., Namba, S. and Mizobuchi, A., "Lattice-site locations of tin and antimony implanted in gallium phosphide", Jpn. J. Appl. Phys., 16 (9), 1647-53 (1977), EEA80-41390.
1211. Takai, M., Ryssel, H., Kranz, H. and Bayerl, P., "Photo- and electroluminescence of nitrogen isoelectronic traps in $GaAs_{1-x}P_x$", Proceedings of the International Conference on Ion Beam Modification of Materials, 1978, p. 1355-66.
1212. Takai, M., Ryssel, H. and Roschenthaler, D., "Nitrogen Implantation in $GaAs_{1-x}P_x$ (X-0.4:0.65)", Appl. Phys., 21 (3), 241-8 (1980), EEA82-29133.
1213. Titley, D. J. and Brown, G., "Thermally stimulated current measurements on proton implanted gallium phosphide", Solid-State Electron, 22 (11), 915-19 (1979), EEA83-9535.
1214. Tremain, R. E., Jr., "Doping gallium arsenide by ion implantation and proton-enhanced diffusion", Thesis, Stanford Univ., Stanford, Calif., 1976, 107 pp., Order No. 76-18807, CA85-152783.
1215. Weyer, G., Petersen, J. W., Damgaard, S., Nielsen, H. L. and Heinemeier, J., "Site-selective doping of compound semiconductors by ion implantation of radioactive nuclei", Phys. Rev. Lett., 44 (3), 155-7 (1980), EEA83-18999.
1216. Wilmsen, C. W. and Kee, R. W., "The improvement of grown oxides for the surface protection of III-V compounds (by ion implantation)", Thin Solid Films, 51 (1), 93-8 (1978), EEA81-42377.
1217. Wolford, D. J., Anderson, R. E. and Streetman, B. G., "Nitrogen implantation in $GaAs_{1-x}P_x$ I. Photoluminescence properties", J. Appl. Phys., 48 (6), 2442-52 (1977), EEA80-32384.
1218. Zilko, J. L., Barnett, S. A., Eltoukhy, A. H. and Greene, J. E., "Modification of elemental incorporation probabilities by ion bombardment during growth of III-V compound and metastable films", J. Vac. Sci. & Technol., 17 (2), 595-602 (1980), EEA83-38728.

4. Group II - VI Semiconductors

1219. Adachi, S. and Machi, Y., "Energy level study of phosphorus-ion-implanted ZnSe", Jap. J. Appl. Phys., 14 (10), 1599-600 (1975), EEA79-835.
1220. Adachi, S. and Machi, Y., "Phosphorus-ion-implantation into ZnSe single crystals", Jap. J. Appl. Phys., 15 (8), 513-21 (1976), EEA79-41396.
1221. Adachi, S. and Machi, Y., "Boron- and fluorine-ion-implantation into ZnSe single crystals", Jpn. J. Appl. Phys., 17 (1), 135-9 (1978), EEA81-14654.
1222. Adachi, S. and Machi, Y., "Anomalous enhancement of electron paramagnetic resonance signal of Mn^{2+} in N-ion-implanted CdS", Jpn. J. Appl. Phys., 17 (1), 229-30 (1978), PA81-27080.

1223. Anon., "MIS diode structure in As^+-implanted CdS", Solid State Technol., 21 (4), 78 (1978), EEA81-33323.

1224. Armitage, S. A., "Radiation-enhanced diffusion of ion-implante bismuth in cadmium sulphide", J. Phys. D, 8 (17), 2034-42 (1975), EEA79-841.

1225. Arkadeva, E. N., Guseva, M. I., Matveev, O. A. and Sladkova, V. A., "Role of lattice defects in the efficiency of implantation doping of cadmium telluride crystals", Sov. Phys.-Semicond., 9 (5), 563-4 (1975), PA79-14716.

1226. Bahir, G., Bernstein, T. and Kalish, R., "Laser annealing of indium-implanted $Pb_{0.8}Sn_{0.2}Te$ films", Appl. Phys. Lett., 34 (8), 486-8 (1979), EEA82-34261.

1227. Bahir, G., Bernstein, T. and Kalish, R., "Damage and lattice location studies in Hg implanted $Hg_{1-x}Cd_xTe$", Radiat. Eff., 48 (1-4), 247-52 (1980), EEA83-53664.

1228. Bean, J. C., "Ion implantation in cadmium telluride", Thesis, Stanford Univ., Stanford, Calif., 1976, 106 pp., Order No. 76-18746, CA85-169847.

1229. Bhar, T. N. and Linder, J. S., "Spectral response characteristics of ion implanted CdS thin films", Indian J. Pure and Appl. Phys., 15 (9), 663-5 (1977), PA81-2512.

1230. Bierce, B. J. and Hengehold, R. L., "Depth-resolved cathodoluminescence of ion-implanted layers in zinc oxide", J. Appl. Phys., 47 (2), 644-51 (1976), PA79-49513.

1231. Bis, R. F. and Houston, B., "Ion implantation of $Pb_{0.8}Sn_{0.2}Te$' IEEE Trans. Nucl. Sci., NS-23 (6), 1546-8 (1979), EEA80-9395.

1232. Bochkov, Y. V., Georgobiani, A. N., Dementev, B. P., Kotlyarevskii, M. B., Noskov, D. A. and Ramazanov, P. E., "Ion-implantation control of stoichiometry of II-VI semiconductors' Sov. Phys.-Semicond., 10 (2), 189-91 (1976), PA79-80468.

1233. Bochkov, Y. V., Ilyukhina, Z. P., Lepnev, L. S. and Tunitskaya, V. F., "Photoluminescence of zinc sulfide with ion-implanted neon impurity", J. Appl. Spectrosc., 30 (2), 179-82 (1979), EEA83-29798.

1234. Bryant, F. J. and Nahum, J., "The effect of non-stoichiometry or the implantation of ytterbium in cadmium telluride", Radiat. Eff., 31 (2), 107-16 (1977).

1235. Bubulac, L. O., Tennant, W. E., Shin, S. H., Wang, C. C., Lanir, M., Gertner, E. R. and Marshall, E. D., "Ion implantation study of HgCdTe", Jpn. J. Appl. Phys., 19, suppl. 19-1, 495-500 (1979), EEA83-48705.

1236. Chu, M., Bube, R. H. and Gibbons, J. F., "Electronic properties of As- and P-implanted cadmium telluride", J. Electrochem. Soc., 127 (2), 483-91 (1980), EEA83-28958.

1237. Chung, C. H. and Kim, K. B., "P-ion implantation into n-type ZnSe", J. Korean Phys. Soc., 7 (2), 105-9 (1974), PA79-27027.

1238. Chung, C. H., Yoon, H. W., Kang, H. S. and Tai, C. H., "N-ion implantation into ZnSe", New Phys., 16 (3), 139-42 (1976), Korean, EEA81-10276.

1239. Fiorito, G., Gasparrini, G. and Svelto, F., "Advances in Hg implanted $Hg_{1-x}Cd_xTe$ photovoltaic detectors", Infrared Phys., 15 (4), 287-93 (1975), EEA79-9263.
1240. Georgobiani, A. N. and Kotljarevsky, M. B., "Ion implantation of zinc sulphide", Radiat. Eff., 47 (1-4), 21-5 (1980), EEA83-44494.
1241. Georgobiani, A. N., Kotlyarevskii, M. B., Lastovka, V. V. and Noskov, D. A., "Production of p-type ZnSe by ion implantation with arsenic", Sov. Phys.-Lebedev Inst. Rep., no. 3, 24-7 (1977), EEA81-45944.
1242. Gresslehner, K. H., Jantsch, W., Lischka, K., Palmetshofer, L. and Heinrich, H., "Ion implantation induced damage and resonant levels in PbTe", Physics of Narrow Gap Semiconductors Symposium, 1977, p. 205-10, EEA83-23922.
1243. Gresslehner, K. H. and Palmetshofer, L., "Ion-implantation-induced damage and resonant levels in $Pb_{1-x}Sn_xTe$", J. Appl. Phys., 51 (9), 4735-8 (1980), PA84-7191.
1244. Jalochowski, M., Mojejko, K., Paprocki, K. and Subotowicz, M., "Recrystallization processes in SnTe thin films implanted with the noble gas ions (Kr, Xe)", Proceedings of the International Conference on Ion Beam Modification of Materials, 1978, p. 977-84.
1245. Katircioglu, B., Pautrat, J. L., Bensahel, D. and Magnea, N., "Dual ion implantation in ZnTe (O+Zn). Interaction between solubility and stoichiometry", Radiat. Eff., 37 (3-4), 183-91 (1978), EEA81-49729.
1246. Koehler, T., "Ion implantation of gold in mercury cadmium telluride", Patent U. S. 4,003,759, Publ. January 1977, CA86-99915.
1247. Kuchar, F., Palmetshofer, L., Ramage, J. C. and Stradling, R. A., "Magneto-optical experiments on ion implanted PbTe with lasers in the far infrared", Acta Phys. Austriaca. Laser Spectroscopy., Suppl. 20, 19-21 (1978), PA82-28848.
1248. Kutra, J., Sakalas, A., Zindulis, A. and Zuk, V., "Electrical properties of p-CdSe layers obtained by the ion implantation of selenium", Thin Solid Films, 55 (3), 421-5 (1978), EEA82-26677.
1249. Kwietniak, M. S., "The luminescence measurements of ZnTe implanted by Li, Cl, Al and In ions", Electron Technol., 10 (2), 79-90 (1977), EEA81-10265.
1250. Ligeon, E., Pautrat, J. L. and Bouriant, M., "Anomalous penetration of implanted ^{65}Zn and compensated zone build-up in ZnTe crystals", Phys. Lett. A, 59A (4), 307-10, (1976), EEA80-9407.
1251. Margalit, S., Menirovsky, Y. and Rotstein, I., "Electrical properties of ion-implanted layers in $Hg_{0.79}Cd_{0.21}Te$", J. Appl. Phys., 50 (10), 6386-9 (1979), EEA83-9529.
1252. Marine, J., Pautrat, J. L., Pfister, J. C., Quillec, M. and Verdone, M., "Ion implantation in II-VI semiconductors", Acta Electron., 19 (2), 161-8 (1976), French, PA79-66111.

1253. Norris, C. B., "Cathodoluminescence studies of anomalous ion implantation defect introduction in ZnTe", J. Appl. Phys., 51 (4), 1998-2002 (1980), EEA83-34801.

1254. Norris, C. B., Barnes, C. E. and Zanio, K. R., "Cathodoluminescence studies of anomalous implantation defect introduction in CdTe", J. Appl. Phys., 48 (4), 1659-67 (1977), EEA80-28822.

1255. Palmetshofer, L., Heinrich, H., Benka, O. and Rescheneder, W., "Ion-implantation-included lattice defects in PbTe", Appl. Phys. Lett., 30 (11), 557-9 (1977), EEA80-32377.

1256. Palmetshofer, L., Vierlinger, E., Heinrich, H. and Haas, L. D., "Evaluation of doping profiles in ion-implanted PbTe", J. Appl. Phys., 49 (3), pt. 1, 1128-30 (1978), EEA81-33268.

1257. Pautrat, J. L., Bensahel, D., Katircioglu, B., Pfister, J. C. and Reviol, L., "Ion implantation in ZnTe: defect generation, migration and annealing", Radiat. Eff., 30 (2), 107-21 (1976), PA79-87919.

1258. Roche, J. P., Dupuy, M. and Lafenille, D., "Native defects, and defects created by boron implantation in zinc telluride", Electrochemical Society Fall Meeting, Extended Abstracts, 1976, p. 889-91, EEA80-36257.

1259. Ryssel, H., Mueller, K., Biersack, J., Krueger, W., Lang, G. and Jahnel, F., "Range and range straggling of ion-implanted boron in $Cd_{0.2}Hg_{0.8}Te$", Phys. Status Solidi A, 57 (2), 619-24 (1980).

1260. Shiraki, Y., Shimada, T. and Komatsubara, K. F., "Edge emissions of ion-implanted CdS", J. Phys. and Chem. Solids, 38 (9), 937-41 (1977), PA80-77274.

1261. Terada, T., "Photo-conductivity of phosphorus-ion-implanted zinc selenide", J. Phys. Soc. Jap., 40 (4), 1048-55 (1976), PA79-58546.

1262. Walsh, D., "Luminescence and crystal damage in ion implanted CdS and ZnO", Solid-State Electron, 20 (10), 813-15 (1977), EEA81-6368.

1263. Yamaguchi, M., "Annealing behaviour of the insulating layer in ion-implanted CdS", Jap. J. Appl. Phys., 15 (4), 723-4 (1976), PA79-54466.

1264. Yu, P. W., Grant, J. T., Park, Y. S. and Haas, T. W., "Zn-ion-implantation profiles in $CuInSe_2$ by Auger electron spectroscopy", J. Appl. Phys., 48 (1), 67-72 (1977), EEA80-20911.

5. Dielectric Thin Films

1265. Arnold, G. W., "Vibrational and electronic spectroscopy of ion implantation-induced defects in fused silica and crystal line quartz", In: Phys. SiO_2 Its Interfaces, Proc. Int. Top. Conf., Pantelides, T., (Ed.), Pergamon, Elmsford, N. Y., p. 278-83 (1978), CA90-78738.

1266. Arnold, G. W., "Controlled surface crystallization of fused silica by Li^+-ion implantation", Radiat. Eff., 47 (1-4), 15-20 (1980), PA83-89383.
1267. Arnold, G. W., "Surface and bulk vibration in ion-implanted amorphous silica", Report SAND-80-0431C, Sandia National Labs., Albuquerque, N. M., 1980, 5 pp.
1268. Arnold, G. W. and Borders, J. A., "Aggregation and migration of ion-implanted silver in lithia-alumina-silica glass", J. Appl. Phys., 48 (4), 1488-96 (1977), PA80-55240.
1269. Astakhov, V. P. and Konyushenko, L. Y., "Crystallization of oxide layers obtained on Si by oxygen-ion implantation", Inorg. Mater., 13 (5), 740-1 (1977), PA81-34724.
1270. Angulis, L., Pranevicius, L. and Seleznev, V. N., "Effect of ion-implanted oxygen on optical properties of SiO films", Sov. Phys.-Collect. 18 (4), 63-6 (1978), PA83-45027.
1271. Badawi, M. H. and Anand, K. V., "A study of silicon oxides prepared by oxygen implantation into silicon", J. Phys. D., 10 (14), 1931-42 (1977), EEA80-41421.
1272. Bayerl, P., Ryssel, H. and Ramin, M., "High energy implantation of buried insulating layers", Proceedings of the International Conference on Ion Beam Modification of Materials, 1978, p. 1187-94.
1273. Belson, J. and Wilson, I. H., "Hydrogen and deuterium ion bombardment effects in SiO_2 films", IEEE Trans. Nucl. Sci., NS-26 (6), 4819-23 (1975), EEA83-18957.
1274. Belson, J. And Wilson, I. H., "Changes in annealing behaviour of SiO_2 on Si resulting from H_2^+ and D_2^+ ion bombardment", Electron Lett., 15 (14), 420-21 (1979), EEA82-38141.
1275. Bitner, L. R., Vedernikov, V. A., Danilina, T. I., Manyakhina, G. V. and Shandra, Z. A., "Mechanism of the effect of electron-ion bombardment during reactive sputtering on the growth of dielectric films", Izv. Vyssh, Uchebn. Zaved., Fiz., 19 (12), 11-15 (1976), CA86-149377.
1276. Bottiger, J., Leslie, J. R. and Rud, N., "Range profiles of 6-16 keV hydrogen ions implanted in metal oxides", J. Appl. Phys., 47 (4), 1672-5 (1976), PA79-68591.
1277. Briska, M., Seaboldt, J. J. and Thiel, K. P., "Reduction of charge in silicon dioxide layers by use of chlorine on ion implantation", IBM Tech. Disclosure Bull., 23 (7), 2768 (1980).
1278. Carnera, A., Drigo, A. V. and Mazzoldi, P., "Atom location in complex lattices: Pb in α-Al_2O_3", Proceedings of the International Conference on Ion Beam Modification of Materials, 1978, p. 297-304.
1279. Christodoulides, C. E., Grant, W. A. and Williams, J. S., "Rutherford backscattering analysis of ion-implanted, thermally oxidized silicon", J. Electronchem. Soc., 124 (10), 1651-3 (1977), EEA81-10286.

1280. DeKeersmaecker, R. F. and DiMaria, D. J., "Electron trapping and detrapping characteristics of arsenic-implanted SiO_2 layers", J. Appl. Phys., 51 (2), 1085-1101 (1980).

1281. DeKeersmaecker, R. F., DiMaria, D. J. and Pantelides, S. T., "Photodepopulation of electrons trapped in silicon dioxide on sites related to arsenic and phosphorus implantation", In: Phys. SiO_2 Its Interfaces, Proc. Int. Top. Conf., Pantelides, S. T., (Ed.), Pergamon, Elmsford, N.Y., 189-94 (1978), CA90-79617.

1282. Dieumegard, D., Croset, M., Cohen, C., Lhoir, A., Rigo, S. and Chaumont, J., "Analysis of arsenic implantation profiles in silica by nuclear microanalysis and secondary ion emission", Vide, 171, Suppl., 263-9 (1974), CA86-64386.

1283. DiMaria, D. J., Young, D. R., DeKeersmaecker, R. F., Hunter, W. R. and Serrano, C. M., "Centroid location of implanted ions in the SiO_2 layer of MOS structures using the photo I-V technique", J. Appl. Phys., 49 (11), 5441-4 (1978), EEA80-12566.

1284. Dylewski, J. and Joshi, M. C., "Thin SiO_2 films formed by oxygen ion implantation in silicon: electron microscope investigations of the Si-SiO_2 interface structures and their C-V characteristics", Thin Solid Films, 37 (2), 241-8 (1976), EEA80-626.

1285. Dylewski, J. and Joshi, M. C., "The dielectric breakdown properties and I-V characteristics of thin SiO_2 films formed by high dose oxygen ion implantation into silicon", Thin Solid Films, 42 (2), 227-35 (1977), EEA80-28818.

1286. Edelman, F. L., Kuznetsov, O. N., Lezheiko, L. V. and Lubopytova, E. V., "Formation of SiC and Si_3N_4 in silicon by ion implantation", Radiat. Eff., 29 (1), 13-15 (1976), EEA79-28435.

1287. EerNisse, E. P., "Stress in ion-implanted CVD Si_3N_4 films", J. Appl. Phys., 48 (8), 3337-41 (1977), EEA81-1184.

1288. Gartner, W. and Schulz, M., "Electronic conduction mechanisms of Cs-and B-implanted SiO_2-films", Appl. Phys., 12 (2), 137-48 (1977), EEA80-9350.

1289. Gavrilenko, V. I., Zuev, V. A. and Litovchenko, V. G., "Influence of the ion implantation on electroreflection spectra of the Si-SiO_2 system", Ukr. Fiz. Zh., 20 (11), 1868-71 (1975), Russian, EEA79-8858.

1290. Gerasimenko, N. N., Kovalevskaya, T. I. and Tseitlin, G. M., "Investigation of hydrogen capture in ion-bombarded silicon dioxide films by the ATR method", J. Appl. Spectrosc, 28 (5), 600-2 (1978).

1291. Gill, S. S. and Wilson, I. H., "Rutherford backscattering analysis of oxide layers formed by ion implantation into single-crystal silicon", Thin Solid Films, 55 (3), 435-48 (1978), EEA82-26765.

1292. Gleaves, G. H., Dearnaley, G. and Collins, R. A., "The effects of ion-implanted impurities on the anodic oxidation of titanium", J. Phys. F., 7 (10), 2031-9 (1977), PA80-90502.
1293. Gopal, V., Rao, M. G., Sarma, N., Bhattacharya, P. K. and Wagh, A. G., "Ion implantation and annealing behavior of thermally grown silicon dioxide films", Proc. Nucl. Phys. Solid State Phys. Symp., 1976, p. 423-5, CA89-35256.
1294. Gupta, S. C., Sharma, B. L. and Agashe, V. V., "Nitrogen ion implantation into SiO_2", Surf. Technol., 10 (2), 153-5 (1980), EEA83-43995.
1295. Haack, D., "Electrical properties of thermally grown SiO_2 layers undoped and doped, by means of ion implantation", 4th International Conference on the Physics of Non-Crystalline Solids., 1976, p. 284-90, EEA81-14585.
1296. Haack, D., "Experiments and model of Poole-Frenkel-current in highly doped amorphous silicon dioxide", Amorphous and Liq. Semicond.: Proc. of the Int. Conf., 7th, 1977, p. 673-77.
1297. Haack, D. and Wagemann, H. G., "Electronic conductivity in highly doped SiO_2 by means of ion implantation", Phys. Status Solidi A, 56 (2), 581-91 (1979), EEA83-18962.
1298. Hartemann, P., "Influence of annealing on the surface-acoustic-wave velocity increase induced by ion implantation in quartz", J. Appl. Phys., 49 (10), 5334-5 (1978), EEA82-8437.
1299. Hartemann, P., Doussineau, P. and Levelut, A., "Surface-acoustic-wave velocity in ion-implanted quartz at very low temperatures", Appl. Phys. Lett., 33 (3), 219-21 (1978), PA81-84040.
1300. Hayashi, T., Maeyama, S. and Yoshii, S., "TEM, AES and XPS studies of Si layer on buried SiO_2 layer formed by high-dose oxygen ion-implantation", Jpn. J. Appl. Phys., 19 (6), 1111-16 (1980), PA83-88172.
1301. Heibei, J. and Voges, E., "Refractive index profiles of ion-implanted fused silica", Phys. Status Solidi A, 57 (2), 609-18 (1980).
1302. Hoffmann, G., Hermann, L., Kovacs, P., Puskas, L. and Nagy, A., "Comparison making use of IR absorption of the boron and phosphorus concentrations of SiO_2 films doped by pyrolysis and ion implantation", J. Phys. D, 10 (11), 1509-19 (1977), PA80-72802.
1303. Igras, E., "The effect of ionic bombardment on removal of conductivity type inversion in interface layer of thermally oxidized silicon", Proceedings of the International Conference on Ion Beam Modification of Materials, 1978, p. 363-70.
1304. Jacobs, E. P. and Dorda, G., "Optically induced charge storage in ion implanted SiO_2", Solid-State Electron., 20 (4), 367-72 (1977), EEA80-20857.

1305. Johnson, N. M., Johnson, W. C. and Lampert, M. A., "Electron trapping in aluminum-implanted silicon dioxide films on silicon", Report AD-A010184/OSL, Princeton Univ., N. J., Dept. of Electrical Engineering, 1976, 9 pp.

1306. Johnson, W. C., "Study of electronic transport and trapping in technologically important insulators", Report AD-A031332/OSL, Princeton Univ., N. J., Dept. of electrical Engineering, 1976, 52 pp.

1307. Kastl, R. H., "Heavy ion penetration into silicon dioxide and silicon nitride", Electrochemical Society Fall Meeting, Extended Abstracts, 1976, p. 865, EEA80-36254.

1308. Katenkamp, U., Karge, H. and Prager, R., "Radiation defects and optical properties of ion implanted silicon dioxide", Radiat. Eff., 48 (1-4), 31-4 (1980), PA83-107293.

1309. Kirov, K. I., Atanasova, E. D., Alexandrova, S. P., Amov, B. G. and Djakov, A. E., "Properties of SiO_2 films formed by oxygen implantation into silicon", Thin Solid Films, 48 (2), 187-92 (1978).

1310. Koenig, W. G., Makris, J. S. and Masters, B. J., "Ion implantation through an electrically insulating layer", Patent German 2531003, Publ. February 1976, CA84-143817.

1311. Komarov, F. F., Rogalevich, I. A. and Tishkov, V. S., "Properties investigation of thin silicon nitride layers synthesized by ion implantation", Radiat. Eff., 39 (3-4), 163-7 (1978), PA82-46974.

1312. Krefft, G. B., Beezhold, W. and EerNisse, E. P., "Effect of ionizing radiation on displacement damage in ion-bombarded single crystal α-Al_2O_3 and α-SiO_2", IEEE Trans. Nucl. Sci., NS-22 (6), 2247-9 (1975), PA79-53180.

1313. Learn, A. J. and Hess, D. W., "Effects of ion implantation on charges in the silicon-silicon dioxide system", J. Appl. Phys. 48 (1), 308-12 (1977), EEA80-20840.

1314. Malin, K. and Schmitt, A., "Method of selectively changing refractive indices of passivating materials by means of ion implantation", IBM Tech. Disclosure Bull., 21 (12), 4883 (1979), EEA83-756.

1315. McCaughan, D. V., Kushner, R. A., Simms, D. L. and White, C. W., "Effects of bombardment by low-energy neutral particles on silicon dioxide films", J. Appl. Phys., 51 (1), 299-304 (1980).

1316. Myers, D. R. and Streetman, B. G., "Detection by low temperature photoluminescence of oxygen recoils in 'through-oxide' arsenic implanted silicon", Solid State Commun., 18 (7), 815-17 (1976), PA79-54473.

1317. Nakano, M. and Togei, R., "Surface potential of SiO_2 during ion implantation", Charge Storage, Charge Transport and Electrostatics with Their Applications, 1978, p. 210-11, PA83-19826.

1318. North, J. C., McGahan, T. E., Adams, A. C. and Rice, D. W., "Tapered windows in phosphorus-doped SiO_2 by ion implantation", IEEE Trans. Electron Devices, ED-25 (7), 809-12 (1978), EEA81-38108.
1319. Ormond, D. W., Irene, E. A., Baglin, J. E. E. and Crowder, B. L., "Expansion of thermally grown silicon dioxide thin films upon irradiation with energetic ions", Ion Implantation Semicond. Proc. Int. Conf. Ion Implantation Semicond. Other Mater., 5th, 1977, p. 305-17, CA89-69571.
1320. Parry, P. D. and Bristol, S. P., "Enhanced etching of ion-implanted silicon nitride in buffered hydroflouric acid", J. Vac. Sci. and Technol., 15 (2), 664-7 (1978), EEA81-42365.
1321. Pavolov, P. V., "Structure of silicon nitride synthesized by nitrogen ion implantation, and the resonance mechanism of the transition from the alpha to the beta form during annealing", Sov. Phys.-Crystallogr., 24 (3), 277-80 (1979), PA83-33981.
1322. Pivot, J., Morelli, D., Cachard, A., Tardy, J., Roger, J. A., Thomas, J. P. and Dupuy, C. H. S., "Oxygen implantation in stoichiometric silicon monoxide", Applications of Ion Beams to Materials, 1975, p. 117-20, Publ. 1976, PA79-68593.
1323. Pranevicius, L., Dudonis, J. and Titas, V., "Effect of an electric field on the formation of dielectric films during ion implantation", Sov. Phys.-Collect., 16 (4), 68-71 (1976), PA81-2801.
1324. Ramin, M., Ryssel, H. and Kranz, H., "Oxidation inhibiting properties of Si_3N_4-layers produced by ion implantation", Appl. Phys., 22 (4), 393-7 (1980), EEA83-53645.
1325. Remizovich, V. S. and Rudenko, A. I., "The effect of space charge electric field on ion implantation in dielectrics", J. Electrostat., 6 (3), 259-71 (1979), PA82-82250.
1326. Rideout, V. L., "Process for providing self aligned doping regions by ion implantation and lift off", Patent USA 4144101, Publ. March 1979.
1327. Royce, B. S. H. and Wang, S. T., "The stability of dielectric coatings under radiation", Report AD-A030670/4SL, Princeton Univ., N. J., Solid State and Materials Lab., 1976, 22 pp.
1328. Runge, H., "Spatial position of deep levels near the $Si\text{-}SiO_2$ interface of ion implanted mos structures", IEEE Trans. Electron Devices, ED-23 (11), 1233-6 (1976), EEA80-6547.
1329. Russell, T. J., Royce, B. S. H. and Harari, E., "Displacement damage and radiation effects in boron implanted sapphire", IEEE Trans. Nucl. Sci., NS-22 (6), 2250-2 (1975), PA79-53236.
1330. Sieber, N., Klabes, R. and Ullrich, H., "Effect of ion implantation on charge carrier transfers in silicon dioxide thin layers", Int. Conf. Ion Implantation Semicond., Proc., 1978, p. 354-62, CA90-131607.
1331. Srour, J. R., Othmer, S., Curtis, O. L. and Chiu, K. Y., "Radiation effects on oxides, semiconductors, and devices", Report AD-A028034/7SL, Northrop Research and Technology Center, Hawthorne, Calif., 1976, 182 pp.

1332. Stein, H. J., "Absorption edge and ion bombardment of silicon nitride", J. Appl. Phys., 47 (8), 3421-6 (1976), PA79-84188.

1333. Topich, J. A., "Effects of ion implanted flourine in silicon dioxide", Cecon 78, The 25th Annual Cleveland Electronics Conference, 1978, p. 70-3, EEA82-12691.

1334. Topich, J. A., "Compensation of mobile-ion movement in SiO_2 by ion implantation", Appl. Phys. Lett., 33 (11), 967-9 (1978), EEA82-12520.

1335. Tsujide, T., Nojiri, M. and Kitagawa, H., "Formation of silicon nitride compound layers by high-dose nitrogen implantation", J. Appl. Phys., 51 (3), 1605-10 (1980), EEA83-34800.

1336. Tsukamoto, K., Akasaka, Y. and Horie, K., "Range distribution of implanted arsenic in silicon dioxide", Jpn. J. Appl. Phys., 16 (4), 663-4 (1977), EEA80-20916.

1337. Ufert, K.-D., "Doping of VO_2 thin films by ion implantation", Phys. Status Solidi A, 42 (1), 187-90 (1977), EEA80-41283.

1338. Vromen, B. H., "Ion implantation into and through silicon dioxide; its influence on oxide integrity", Electrochemical Society Fall Meeting (Extended Abstracts), 1975, p. 341-2, EEA79-16894.

1339. Wach, W. and Wittmaack, K., "Comparison of xenon retention in ion implanted silicon dioxide and oxygen-doped silicon", Nucl. Instrum. and Methods, 149 (1-3), 259-64 (1977), PA82-4320.

1340. Wada, Y. and Sato, K., "Experimental determination of current paths of ions implanted into insulators", Jap. J. Appl. Phys., 15 (11), 2289-90 (1976), EEA80-6702.

1341. Webb, A. P., Houghton, A. J. and Townsend, P. D., "Changes in the chemical stability of ion-implanted silica glass", Radiat. Eff., 30 (3), 177-82 (1976), EEA80-1158.

1342. Webb, A. P. and Townsend, P. D., "Refractive index profiles induced by ion implantation into silica", J. Phys. D, 9 (9), 1343-54 (1976), CA85-38818.

1343. Wild, W., Glaefeke, H. and Fitting, H. J., "Investigations of implantation and damage profiles in the silicon dioxide layers of silicon dioxide-silicon-systems using TSEE technique", Proc. Int. Symp. Exoelectron Emiss. Dosim., 5th, 1976, p. 200-2, CA88-112288.

1344. Wolkenberg, A. and Wasilewska, B., "Properties of boron implanted silicon dioxide", Appl. Surf. Sci., 3 (1), 83-7 (1979), EEA82-34297.

1345. Yadav, A. D. and Joshi, M. C., "Formation of thin Si_3N_4 films by nitrogen ion implantation into silicon", Thin Solid Films, 59 (3), 313-17 (1979), PA82-99455.

1346. Young, D. R., DiMaria, D. J. and Hunter, W. R., "The electron trapping behavior of silicon dioxide with ion implanted aluminum", J. Electron. Mater., 6 (5), 569-79 (1977), EEA81-38090.

1347. Young, D. R., DiMaria, D. J., Hunter, W. R. and Serrano, C. M., "Characterization of electron traps in aluminum-implanted SiO_2", IBM J. Res. & Dev., 22 (3), 285-8 (1978), EEA82-8233.
1348. Young, V. Y., Gibbs, R. A., Winograd, N. and Kim, K. S., "X-ray photoemission studies of atom implanted solids: silver and gold in silicon dioxide", Chem. Phys. Lett., 54, 378-82 (1978), CA89-35183.
1349. Zhizhin, G. N., Yakovlev, V. A. and Shirmer, G., "Effect of ion implantation on the reflection and surface polaritons of crystalline quartz", Pis'ma V Zh. Eksp. and Teor. Fiz., 29 (6), 350-3 (1979).

6. Metals and Alloys

1350. Ali, A., Grant, W. A. and Grundy, P. J., "Amorphous Ni-P alloys prepared by ion implantation", Radiat. Eff., 34 (4), 251-4 (1977), PA81-22748.
1351. Ali, A., Grant, W. A. and Grundy, P. J., "A microstructural study of some amorphous transition metal-metalloid surface alloys formed by ion implantation", Philos. Mag. B, 37 (3), 353-76 (1978), PA81-67730.
1352. Al-Saffar, A. H., Ashworth, V., Bairamov, A. K. O., Chivers, D. J., Grant, W. A. and Procter, R. P. M., "Effect of molybdenum ion implantation on the general and pitting corrosion behaviour of pure aluminium and high strength aluminium alloy", Corros. Sci., 20 (1), 127-44 (1980).
1353. Al-Saffar, A. H., Ashworth, V., Grant. W. A. and Procter, R. P. M., "The role of mercury in the dissolution of aluminum sacrificial anodes: a study using ion implantation", Corros. Sci., 18 (8), 687-700 (1978), PA82-39016.
1354. Ashworth, V., Baxter, D., Grant, W. A. and Procter, R. P. M., "The effect of ion implantation on the corrosion behaviour of pure iron. III. Tantalum ion implantation", Corros. Sci., 17 (12), 947-60 (1977), PA81-31559.
1355. Ashworth, V., Grant, W. A. and Procter, R. P. M., "The effect of ion implantation on the aqueous corrosion of metals", Proceedings of the International Conference on Ion Beam Modification of Materials, 1978, p. 1381-2.
1356. Ashworth, V., Grant, W. A. and Procter, R. P. M., "Novel corrosion resistant alloys by ion implantation", Ind. and Eng. Chem. Prod. Res. and Dev., 17 (2), 176-82 (1978), PA81-88569.
1357. Ashworth, V. Grant, W. A., Procter, R. P. M. and Wright, E. J., "The effect of ion implantation on the corrosion behaviour of pure iron. IV. Lead ion implantation", Corros. Sci., 18 (8), 681-5 (1978), PA82-39015.

1358. Asundi, V. K., Joshi, M. C., Deb, S. K., Kulkarni, V. N., Sood, D. K. and Sundararaman, M., "Thermal migration of iron implanted in aluminium at high doses", Proceedings of the International Conference on Ion Beam Modification of Materials, 1978, p. 1815-26.

1359. Bauriedl, W. and Heim, G., "Superconducting properties of ion implanted InMn alloys", Z. Phys. B, 26 (1), 29-36 (1977), PA80-15987.

1360. Beliy, I. M. and Komarov, F. F., "Structural transformations and electro-physical properties of metal films bombarded with ions", Proceedings of the International Conference on Ion Beam Modification of Materials, 1978, p. 1903-12.

1361. Bennett, M. J., Dearnaley, G., Houlton, M. R., Hawes, R. W. M., Goode, P. D. and Wilkins, M. A., "The influence of surface ion implantation upon the oxidation behaviour of a 20 percent Cr-25 percent Ni, niobium stabilized austenitic stainless steel, in carbon dioxide, at 825°C", Corros. Sci., 20 (1), 73-89 (1980).

1362. Bennett, M. J., Houlton, M. R. and Dearnaley, G., "Influence of the surface ion implantation of aluminium and yttrium upon the oxidation behaviour of a Fe-15% Cr-4% Al ferralloy stainless steel, in air, at 1100°C", Corros. Sci., 20 (1), 69-72 (1980).

1363. Borders, J. A., "Metastable phases produced by ion implantation in metals", Report SAND-77-0700C, Sandia Labs., Albuquerque, N. Mex., 1977, 8 pp.

1364. Bottiger, J., Picraux, S. T., Rud, N. and Laursen, T., "Trapping of hydrogen isotopes in molybdenum and niobium predamaged by ion implantation", J. Appl. Phys., 48 (3), 920-6 (1977), PA80-47622.

1365. Campbell, A. B., Sartwell, B. D. and Needham, P. B., Jr., "Composition depth profiles and the effects of annealing for ion-implanted alloys", Report PB80-122765, Bureau of Mines, Avondale, MD, 1979, 31 pp.

1366. Campbell, A. B., Sartwell, B. D. and Needham, P. B., Jr., "Depth profiling of ion-implanted alloys", J. Appl. Phys., 51 (1), 283-9 (1980).

1367. Chadderton, L. T., Johnson, E. and Komolov, S. A., "On the annealing behaviour of dysprosium ion implanted nickel: a combined study using Rutherford backscattering, transmission electron microscopy, and total current spectroscopy", Nucl. Instrum. and Methods, 149 (1-3), 639-45 (1978), PA82-4343.

1368. Chakrabortty, S. B., Spooner, S. and Starke, E. A., "The effect of ion plating and ion implantation on the cyclic response and fatigue crack initiation of metals and alloys", Report AD-A083643/7, Georgia Inst. of Tech., Atlanta, 1980, 59 pp.

1369. Chowdhury, S. K. R., Hartley, N. E. W., Pollock, H. M. and Wilkins, M. A., "Adhesion energies at a metal interface: the effects of surface treatments and ion implantation", J. Phys. D, 13 (9), 1761-84 (1980), PA83-107660.
1370. Covino, B. S., Jr,. Needham, P. B., Jr. and Conner, G. R., "Anodic polarization behavior of Fe-Ni alloys fabricated by ion implantation", J. Electrochem. Soc., 125 (3), 370-2 (1978), PA81-48562.
1371. Covino, B. S., Jr., Sartwell, B. D. and Needham, P. B., Jr., "Anodic polarization behavior of Fe-Cr surface alloys formed by ion implantation", J. Electrochem. Soc., 125 (3), 366-9 (1978), PA81-48561.
1372. Christie, W. H., Carter, J. A., Eby, R. E., Landau, L. and Musick, W. R., "SSMS near surface analysis of B in irradiated zircaloy-2: ion implantation standards as a calibration technique", Report CONF-800568-1, Oak Ridge National Lab., TN, 1980, 4 pp.
1373. Cullis, A. G., Borders, J. A., Hirvonen, J. K., Poate, J. M. and Misell, D. L., "Combined transmission electron microscope and ion channeling study of metastable metal alloys formed by ion implantation", Developments in Electron Microscopy and Analysis, 1977, p. 181-4, PA81-34530.
1374. Dearnaley, G., "Ion implantation of metals. II", Electroplat. and Met. Finish., 29 (7-8), 27-9 (1976), EEA80-2421.
1375. Dearnaley, G., "The ion implantation of metals and engineering materials", Trans. Inst. Met. Finish., 56, Pt. 1, 25-31 (1978), PA82-34204.
1376. Dearnaley, G. and Hartley, N. E. W., "Ion implantation into metals and carbides", Thin Solid Films, 54 (2), 215-32 (1978), PA82-28008.
1377. Duckworth, R. G. and Wilson, I. H., "Ion bombardment synthesis of thin film transition metal nitrides", Proceedings of the International Conference on Ion Beam Modification of Materials, 1978, p. 1207-30.
1378. Duesbury, M. S. and Taylor, R., "Some properties of ion-implanted Li in Be", J. Phys. F, 9 (2), L19-22 (1979), PA82-37876.
1379. Echt, O., Recknagel, E., Weidinger, A. and Wichert, T., "Point defect trapping at ^{111}In in copper after ion implantation", Hyperfine Interactions, 4 (1-2), 706-9 (1978), PA81-43728.
1380. Engman, U. and Holmqvist, B., "Electron microscopy studies of void swelling and annealing of voids in aluminium irradiated with aluminium ions", Proceedings of the International Conference on Ion Beam Modification of Materials, 1978, p. 1913-22.
1381. Ferber, H., Kasten, H., Wolf, G. K., Lorenz, W. J., Schwikert, H., and Folger, H., "Influence of ion implantation on the corrosion behavior of iron in acid solution", Corros. Sci., 20 (1), 117-25 (1980).

1382. Follstaedt, D. M. and Myers, S. M., "Microstructure of Fe implanted with Ti", Proceedings of the International Conference on Ion Beam Modification of Materials, 1978, p. 1839-48.

1383. Gangulee, A., Dheurle, F. M., Ranieri, V. A. and Fiorio, R. A., "Effects of oxygen ion implantation on the structure and properties of aluminium thin films", J. Vac. Sci. and Technol., 16 (2), 156-9 (1979), PA82-77936.

1384. Grant, W. A., "Amorphous metals and ion implantation", J. Vac. Sci. and Technol., 15 (5), 1644-9 (1978), PA82-14686.

1385. Grant, W. A., Chadderton, L. T., Johnson, E., Ali, A. and Grundy, P. J., "Production of amorphous alloys by ion implantation", Symposium on Rapidly quenched Metals III, Part I, Published by Metals Society, London, England, 1978, p. 63-72, PA83-75257.

1386. Grant, W. A., Whitton, J. L. and Williams, J. S., "Ion bombardment induced amorphous layers on single crystal stainless steel - a combined Rutherford backscattering and transmission electron microscopy study", Proceedings of the International Conference on Ion Beam Modification of Materials, 1978, p. 1883-92.

1387. Grundy, P. J., Ali, A., Christodoulides, C. E. and Grant, W. A., "Amorphous transition metal-metalloid alloy films prepared by ion implantation", Thin Solid Films, 58 (2), 253-8 (1979), PA82-91272.

1388. Gumanski, G. A. and Chodasewitsch, V. V., "Investigation of $1s_{1/2}$O, $1s_{1/2}$C and $2p_{3/2}$Ni states of argon ion implanted nickel metal", Krist. and Tech., 14 (5), 581-8 (1979), PA83-16149.

1389. Heim, G., Bauriedl, W. and Buckel, W., "Towards amorphization of indium by ion implantation", J. Nucl. Mater., 72 (1-2), 263-9 (1978), PA81-52135.

1390. Hertel, B., Kammholz, K. and Diehl, J., "Sub-surface damage of metals after 5 keV argon-ion bombardment", Proceedings of the International Conference on Ion Beam Modification of Materials, 1978, p. 1827-38.

1391. Hikosaka, K., Ishiwara, H. and Furukawa, S., "Channeled ion implantation through metallic films", J. Vac. Sci. & Technol., 16 (6), 1913-16 (1979), EEA83-34807.

1392. Hirvonen, J. K., "Ion implantation in tribology and corrosion science", J. Vac. Sci. & Technol., 15 (5), 1662-8 (1978), PA82-14830.

1393. Hirvonen, J. K., "Implantation into metals - mechanical property changes", Proceedings of the International Conference on Ion Beam Modification of Materials, 1978, p. 1753-70.

1394. Hirvonen, J. K., Butler, J. W., Smith, T. P., Kant, R. A. and Westcott, V. C., "Sliding-wear reduction by ion implantation", Proceedings of the International Conference on Ion Beam Modification of Materials, 1978, p. 1973-6.

1395. Hirvonen, J. K., Carosella, C. A., Kant, R. A., Singer, I., Vardiman, R. and Rath, B. B., "Improvement of metal properties by ion implantation", Thin Solid Films, 63 (1), 5-10 (1979), PA83-16601.
1396. Hirvonen, J. K., Poate, J. M., Greenwald, A. and Little, R., Pulsed electron beam irradiation of ion-implanted copper single crystals", Appl. Phys. Lett., 36 (7), 564-6 (1980), PA83-65019.
1397. Hitzfeld, M. and Heim, G., "Influence of oxygen and lattice disorder on the kondo resistivity of ion implanted InMn alloys", Solid State Commun., 29 (2), 93-8 (1979), PA82-38061.
1398. Hu, W. W., Clayton, C. R., Herman, H. and Hirvonen, J. K., "Fatigue-life enhancement by ion implantation", Scr. Metall., 12 (8), 697-8 (1978), PA82-1910.
1399. Hu, W. W., Clayton, C. R., Herman, H., Hirvonen, J. K. and Kant, R. A., "Fatigue-life enhancement of steel by nitrogen implantation", Proceedings of the International Conference on Ion Beam Modification of Materials, 1978, p. 1977-80.
1400. Iwaki, M., Hayashi, H., Kohno, A., Namba, S., Yoshida, K. and Soda, N., "Friction property of chromium, nickel and copper implanted steel", Proceedings of the International Conference on Ion Beam Modification of Materials, 1978, p. 1981-92.
1401. Johnson, E., Wohlenberg, T. and Grant, W. A., "Crystalline phase transitions produced by ion implantation", Phase Transitions, 1 (1), 23-34 (1979), PA83-7558.
1402. Johnson, E., Wohlenberg, T., Grant. W. A., Hansen, P. and Chadderton, L. T., "TEM study of amorphous alloys produced by ion implantation", J. Mircrosc., 116, pt. 1, 77-87 (1979), PA83-15128.
1403. Jones, W. R. and Ferrante, J., "The effect of nitrogen ion implantation on the friction and wear characteristics of iron", Report N79-12201, NASA, Cleveland, Ohio, 1978, 18 pp.
1404. Kaletta, D., "Light element implantations in metals", Proceedings of the International Conference on Ion Beam Modification of Materials, 1978, p. 1571-1626.
1405. Kant, R. A., Hirvonen, J. K., Knudson, A. R. and Wollam, J. S., "Surface hardening of beryllium by ion implantation", Thin Solid Films, 63 (1), 27-30 (1979), PA83-16604.
1406. Kant, R. A., Myers, S. M. and Picraux, S. T., "Experimental study of precipitation in an ion-implanted metal: Sb in Al", J. Appl. Phys., 50 (1), 214-22 (1979), PA82-55766.
1407. Kaufmann, E. N., "Systematics of metastable alloys produced by ion implantation", IEEE Trans. Nucl. Sci., NS-26 (1), pt. 2, 1668-9 (1979), PA82-51132.
1408. Kaufmann, E. N., Aspnes, D. E., Rodgers, J. W. and Studna, A. A., "Dielectric function measurements on ion-implanted metallic systems", J. Opt. Soc. Am., 67 (8), 1099-1101 (1977), PA81-6697.

1409. Kiriakidis, G., Christodoulides, C. E., Carter, G. and Colligon, J. S., "RBS technique for measurement of the erosion rate of ion implanted films", Appl. Phys., 19 (2), 191-4 (1979).

1410. Klugmann, E., "Effect of ion implantation on the magnetic properties of ferromagnetic metals", Phys. Status Solidi A, 39 (1), K53-7 (1977), PA80-28414.

1411. Klugmann, E., "Effects of deformation and ion implantation on the reversible magnetic permeability of nickel", Z. Metallkd., 68 (4), 270-2 (1977), PA80-4813.

1412. Linker, G., "Channeling analysis of radiation disorder in ion implanted vanadium and molybdenum single crystals", Nucl. Instrum. and Methods, 149 (1-3), 365-9 (1978), PA82-4360.

1413. Litovchenko, V. G., Zuev, V. A., Korbutyak, D. V. and Gavrilenko, V. I., "Optical and luminescent properties of ion-implanted films", Thin Solid Films, 66 (3), 255-9 (1980), PA83-60699.

1414. Marwick, A. D. and Piller, R. C., "Modification of implant profiles in nickel by radiation-enhanced diffusion and segregation", Proceedings of the International Conference on Ion Beam Modification of Materials, 1978, p. 1849-62.

1415. Meunier, F., Pfeuty, P., Lamoise, A. M., Chaumont, J., Bernas, H. and Cohen, C., "Percolation and superconductivity in ion-implanted aluminium films", J. Phys. Lett., 38 (21), 435-7 (1977), PA81-26982.

1416. Meyer, J. D. and Stritzker, B., "Investigation of metastable superconducting alloys produced by low temperature ion implantation", Radiat. Eff., 47 (1-4), 177-81 (1980), PA83-88670.

1417. Mizoguchi, T., Gambino, R. J., Hammer, W. N. and Cuomo, J. J., "Effects of ion implantation damage on the magnetic properties of amorphous GdCoMo films", IEEE Trans. Magn., MAG-13 (5), 1618-20 (1977), EEA81-1458.

1418. Morris, J. R., Collins, R. A. and Dearnaley, G., "The influence of ion implantation on the thermal oxidation of copper", J. Phys. F., 8 (6), 1333-42 (1978), PA81-64768.

1419. Myers, S. M., "Annealing behavior and selected applications of ion-implanted alloys", J. Vac. Sci. and Technol., 15 (5), 1650-5 (1978), PA82-14829.

1420. Myers, S. M., "Implanted and annealed alloys in physical metallurgy", Proceedings of the International Conference on Ion Beam Modification of Materials, 1978, p. 1771-96.

1421. Myers, S. M., "Properties and applications of ion-implanted alloys", J. Vac. Sci. & Technol., 17 (1), 310-14 (1980), PA83-79824.

1422. Myers, S. M., Picraux, S. T. and Stoltz, R. E., "Deuterium trapping in iron alloys formed by ion implantation", Thin Solid Films, 63 (1), 37-8 (1979), PA83-15131.

1423. Myers, S. M., Picraux, S. T. and Stoltz, R. E., "Defect trapping of ion-implanted deuterium in Fe", J. Appl. Phys., 50 (9), 5710-19 (1979), PA83-2047.

1424. Myers, S. M., Poate, J. M., Tu, K. N. and Mayer, J. W., "Ion-implanted metal layers", In: Thin Films. Interdiffusion and Reactions, Wiley, New York, N. Y., p. 533-66 (1978), PA82-1835.
1425. Okabe, Y., Iwaki, M., Takahashi, K., Hayashi, H., Namba, S. and Yoshida, K., "Electrochemical properties of ion implanted steel", Surf. Sci., 86 (2), 257-63 (1979), PA83-11840.
1426. Picraux, S. T., Myers, S. M. and Follstaedt, D. M., "Microalloy layer formation by ion implantation", Thin Solid Films, 63 (1), 1-2 (1979), PA83-16600.
1427. Picraux, S. T., Rimini, E., Foti, G. and Campisano, S. U., "Dechanneling by dislocations in ion-implanted Al", Phys. Rev. B, 18 (5), 2078-96 (1978), PA82-7642.
1428. Poate, J. M., "New developments in ion implantation in metals, superconductors, and other materials", J. Vac. Sci. and Technol., 14 (1), 529 (1977), PA80-59027.
1429. Poate, J. M., "Implantation into metals for metastable alloy formation", Proceedings of the International Conference on Ion Beam Modification of Materials, 1978, p. 1797-814.
1430. Poate, J. M., Borders, J. A. and Cullis, A. G., "Ion implantation as an ultrafast quenching technique for metastable alloy production: the Ag-Cu system", Appl. Phys. Lett., 30 (8), 365-8 (1977), PA80-55902.
1431. Proto, G. R. and Lawless, K. R., "Effect of He ion implantation on the domain structure of evaporated single-crystal Ni thin films viewed by Lorentz electron microscopy", Appl. Phys. Lett., 29 (12), 819-22 (1976), EEA80-9572.
1432. Rechtin, M. D., Baldo, P. M. and Van der Sande, J., "Amorphous $^{40}Ni_{60}$ produced by Ni^+ ion implantation", Electron Microscopy 1978, Ninth International Congress on Electron Microscopy, Part I, 1978, p. 388-9, PA83-107062.
1433. Reintsema, S. R., Drentje, S. A. and DeWaard, H., "Mossbauer effect studies of the recovery of iron after heavy-ion implantation at 7°K", Hyperfine Interactions, 5 (2), 167-78 (1978), PA81-52393.
1434. Reintsema, S. R., Odeurs, J., Verbiest, E., Pattyn, H. and Coussement, R., "Search for the vacancy annealing stage in heavy ion implanted BCC metals", Hyperfine Interactions, 4 (1-2), 685-8 (1978), PA81-39463.
1435. Reintsema, S. R., Verbiest, E., Odeurs, J. and Pattyn, H. "Vacancy recovery in heavy ion-implanted refractory BCC metals studied by Mossbauer spectroscopy", J. Phys. F., 9 (8), 1511-27 (1979), PA82-82896.
1436. Rossler, K. and Pross, L., "Molecular units in ionic crystals via abstraction of matrix atoms by implanted metal ions", Proceedings of the International Conference on Ion Beam Modification of Materials, 1978, p. 1451-64.

1437. Ruault, M. O., Bernas, H. and Chaumont, J., "Transmission electron microscopy study of damage by ion implantation in gold: evidence for a spike threshold", Philos. Mag. A., 39 (6), 757-83 (1979), PA82-90705.
1438. Russo, S. L., Mazzoldi, P., Scotoni, I., Tosello, C. and Tosto, S., "Effect of nitrogen-ion implantation on the unlubricated sliding wear of steel", Appl. Phys. Lett., 34 (10), 527-9 (1979), PA82-74261.
1439. Russo, S. L., Mazzoldi, P., Scotoni, I., Tosello, C. and Tosto, S., "Fatigue-life improvement by nitrogen-ion implantation on steel: dose dependence", Appl. Phys. Lett., 36 (10), 822-3 (1980), PA83-76328.
1440. Sartwell, B. D., "Formation of corrosion-resistant surface alloys by metal ion implantation", Thin Solid Films, 54 (2), 233-42 (1978), PA82-28009.
1441. Sartwell, B. D., Campbell, A. B., III, Covino, B. S., Jr. and Driscoll, T. J., "Applications of ion implantation to metallic corrosion", IEEE Trans. Nucl. Sci., NS-26 (1), pt. 2, 1670-6 (1979), PA82-52447.
1442. Sawicka, B. D., "Hyperfine interactions of ^{57}Fe in iron beam modified metals", Proceedings of the International Conference on Ion Beam Modification of Materials, 1978, p. 1893-1902.
1443. Singer, I. L. and Murday, J. S., "Chemical state of ion-implanted nitrogen in Fe18Cr8Ni steel", J. Vac. Sci. & Technol., 17 (1), 327-9 (1980), PA83-79825.
1444. Sleeswyk, A. W., Kok, H. J. G. and Boom, G., "Effect of C-ion implantation on L.C.F. life in Cu, Ni and AISI 310", Scr. Metall., 14 (8), 919-22 (1980), PA84-3920.
1445. Smith, H. J. and Van Wyk, G. N., "Effect of Cu^+ ion implantation on crystalline orientation in polycrystalline Cu investigated by Kr penetration measurements", Radiat. Eff., 46 (3-4), 149-55 (1980), PA83-88174.
1446. Sood, D. K., "Empirical rules for substitutionality in metastable surface alloys produced by ion implantation", Phys. Lett. Vol. A, 68A (5-6), 469-72 (1978), PA82-14832.
1447. Sood, D. K. and Dearnaley, G., "Ion implanted surface alloys in nickel", Radiat. Eff., 39 (3-4), 157-62 (1978), PA82-46692.
1448. Spooner, S., Solnick-Legg, H. and Starke, E. A. "The effect of ion plating and ion implantation on the cyclic response and fatigue crack initiation of metals and alloys", Report AD-A066547, Georgia Inst. of Tech., Atlanta, Georgia, 1979, 29 pp.
1449. Stesmans, A., Van Meijel, J. and Witters, J., "Influence of ion implantation on CeSr in aluminium", J. Phys. F., 7 (1), 199-206 (1977), PA80-20701.
1450. Suri, A. K., Nimmagadda, R. and Bunshah, R. F., "Influence of ion implantation and overlay coatings on various physico-mechanical and wear properties of stainless steel, titanium and aluminium", Thin Solid Films, 64 (2), 191-203 (1979), PA83-40603.

1451. Svendsen, L. G., "A comparison of corrosion protection of copper by ion implantation of Al and Cr", Corros. Sci., 20 (1), 63-8 (1980), PA83-71789.
1452. Wampler, W. R., Follstaedt, D. M. and Picraux, S. T., "Electron beam annealing of ion implanted Al", Appl. Phys. Lett., 36 (5), 366-8 (1980), PA83-59500.
1453. Wang, Y. F., Clayton, C. R., Hubler, G. K., Lucke, W. H. and Hirvonen, J. K., "Applications of ion implantation for the improvement of localized corrosion resistance of M50 bearing steel", Thin Solid Films, 63 (1), 11-18 (1979), PA83-16602.
1454. Whitmell, D. S., Nelson, R. S., Williamson, R. and Smith, M. J. S., "Immobilization of krypton by incorporation into a metallic matrix by combined ion implantation and sputtering", Nucl. Energy, 18 (5), 349-52 (1979), PA83-27794.
1455. Yadava, R. D. S., Ibobi, S. N. and Nigam, A. K., "Sponge-like blisters on copper by H^+-ion implantation at ambient temperatures", J. Phys. D., 13 (11), 2077-80 (1980), PA84-7216.
1456. Yamagishi, H. and Nittono, O., "X-ray study on lattice defects in Ar^+ ion implanted copper whiskers", J. Jpn. Inst. Met., 43 (8), 689-95 (1979), Japanese, PA82-99230.
1457. Zamanzadeh, M., Allam, A., Pickering, H. W. and Hubler, G. K., "Effect of helium-, iron-, and platinum- implantation on the permeation of hydrogen through iron membranes", Report AD-A071595/3SL, Pennsylvania State Univ., University Park, 1979, 28 pp.
1458. Ziemann, P., Heim, G. and Buckel, W., "Low temperature self-ion implantation into granular aluminium films", Radiat. Eff., 48 (1-4), 75-80 (1980), PA83-103774.

7. Others

1459. Allen, W. N., Brant, P., Carosella, C. A., DeCorpo, J. J., Ewing, C. T., Saalfield, F. and Weber, D. C., "Ion implantation studies of $(SN)_x$ and $(CH)_x$", Synth. Met., 1 (2), 151-9 (1980), EEA83-29028.
1460. Arrington, C. H., III, "Superconducting weak links formed by ion implantation", Thesis, Univerity of Virginia, 1975, 82 pp., Order no. 76-18086.
1461. Bauriedl. W., Ziemann, P. and Heim, G., "Ion implantation as a tool for studying the Kondo-system ZnMn", Z. Phys. B, 38 (1), 59-63 (1980), PA83-70241.
1462. Bentini, G. G., Berti, M., Carnera, A., Della Mea, G., Drigo, A. V., Lo Russo, S., Mazzoldi, P. and Dearnaley, G., "Thermal oxidation of ion-implanted zirconium", Corros. Sci., 20 (1), 27-39 (1980).
1463. Cambon, G. and Hartemann, P., "Modelling of ion implanted lithium niobate used for propagating surface acoustic waves", Rev. Phys. Appl., 13 (5), 241-5 (1978), French, EEA81-28688.

1464. Clapp, M. T. and Rose, R. M., "On the synthesis of A-15 'Nb_3Si' by ion implantation", Appl. Phys. Lett., 33 (2), 205-7 (1978), EEA81-46198.
1465. Clapp, M. T. and Rose, R. M., "On the superconducting transition temperature of A-15 'Nb_3Si' synthesized by ion implantation", IEEE Trans. Magn., MAG-15 (1), 772-3 (1979), EEA82-22501.
1466. Cooper, C. M., Nayar, P. S., Hale, E. B. and Gerson, R., "Conductive strontium titanate layers produced by boron-ion implantation", J. Appl. Phys., 50 (4), 2826-31 (1979), EEA82-42401.
1467. Davenas, J., Perez, A., Dupin, J. P. and Dupuy, C. H. S., "Implantation effects in LiF bombarded with alkali ions", Proceedings of the International Conference on Ion Beam Modification of Materials, 1978, p. 1303-12.
1468. Destefanis, G. L., Gailliard, J. P. and Townsend, P. D., "Optical waveguides in $LiNbO_3$ formed by ion implantation", Proceedings of the International Conference on Ion Beam Modification of Materials, 1978, p. 1313-20.
1469. Destefanis, G. L., Townsend, P. D. and GAilliard, J. P., "Optical waveguides in $LiNbO_3$ formed by ion implantation of helium", Appl. Phys. Lett., 32 (5), 293-4 (1978), EEA81-24255.
1470. Diaine, C., Gisclon, J. L. and Dupuy, J., "Colloid centres in AgCl induced by heavy ion implantation", Phys. Status Solidi B., 87 (1), 293-300 (1978), PA81-63780.
1471. Dravin, V. A., Motulevich, G. P. and Anshukova, N. V., "Superconducting properties of the system Nb-N produced by ion implantation", Sb. Kratk. Soobshch. Fiz. An SSR Fiz. Inst. P.N. Lebedev., no. 1, 12-16 (1977), Russian, EEA81-10549.
1472. Fogel, N. Y., Moshenski, A. A. and Dmitrenko, I. M., "Ion implantation as a method of studying inhomogeneities in superconductors: results for indium films with embedded helium particles", J. Nucl. Mater., 72 (1-2), 282-90 (1978), PA81-52181.
1473. Gamo, K., Goshi, H., Takai, M., Iwaki, M., Masuda, K. and Namba, S., "Control of T_c for niobium by N ion implantation", Jpn. J. Appl. Phys., 16 (10), 1853-7 (1977), EEA81-1507.
1474. Gamo, K., Kajiya, A. and Namba, S., "Fabrication of superconducting weak links by electron beam lithography and ion implantation", Jpn. J. Appl. Phys., 17 (4), 731-2 (1978), EEA81-28805.
1475. Gamo, K. and Namba, S., "Josephson junction fabrication by means of ion implantation and electron beam lithography", Proceedings of the International Conference on Ion Beam Modification of Materials, 1978, p. 1993-2004.
1476. Geerk, J., "The superconducting transition temperature of niobium carbide single crystals after implantation of light elements", Proceedings of the International Conference on Ion Beam Modification of Materials, 1978, p. 2035-8.

1477. Gerson, R., "Ion implantation in Perovskite type ferroelectrics", Report AD-A080078/9, Missouri University - Rolla, 1979, 15 pp.
1478. Harris, E. P., "Transition from Josephson behavior to depairing behavior in ion-implanted molybdenum weak links", Solid State Commun., 20 (5), 449-51 (1976), EEA80-756.
1479. Harris, E. P., Henkels, W. H. and Kircher, C. J., "Applications of ion implantation to the Josephson tunneling technology", IBM Tech. Disclosure Bull., 20 (6), 2435-40 (1977), EEA82-4592.
1480. Hartemann, P. and Cauvard, P., "Ion implantation and its effects on the propagation of elastic surface waves", Report N79-27368/6SL, Laboratoire Central de Recherches Thomson-CSF, (France), 1977, 79 pp., French.
1481. Kass, W. J., "Thermal desorption measurements of helium ion-implanted erbium tritide", J. Vac. Sci. and Technol., 14 (1), 518-22 (1977), PA80-60102.
1482. Land, C. E. and Peercy, P. S., "Photosensitivity enhancement by H- and He-ion implantation in lead lanthanum zirconate titanate ceramics", Appl. Phys. Lett., 37 (1), 39-41 (1980), EEA83-49196.
1483. Langguth, K. G., "Production of superconducting vanadium carbide by ion implantation", Report KFK-2476, Kernforschungszentrum Karlsruhe (West Germany), 1977, 56 pp.
1484. Linker, G., "Superconducting properties and structure of ion bombarded Nb layers", Proceedings of the International Conference on Ion Beam Modification of Materials, 1978, p. 2039-48.
1485. Meyer, O., "Ion implantation in superconductors", Proceedings of the International Conference on Ion Beam Modification of Materials, 1978, p. 2005-34.
1486. Okuyama, Y., Hashimoto, T. and Koguchi, T., "High dose ion implantation into photoresist", J. Electrochem. Soc., 125 (8), 1293-8 (1978), EEA82-788.
1487. Rechtin, M. D., Baldo, P. M. and Vander Sande, J., "Ion-implantation damage in amorphous and crystalline $Nb_{40}Ni_{60}$", Scr. Metall., 12 (7), 639-43 (1978), PA81-91365.
1488. Robic, J. Y., Piaguet, J., Duret, D., Veler, J. C., Veran, J. L. and Zenatti, D., "Ion implantation in superconducting niobium and Nb_3Sn thin films: adjustment of Josephson microbridges and SQUID", Proceedings of the International Conference on Ion Beam Modification of Materials, 1978, p. 2061-70.
1489. Sawicki, J. A. and Sawicka, B. D., "Comparability of $_{57}Fe$ Mossbauer data in ion-implanted and amorphous FeGe", Phys. Status Solidi B., 80 (1), K41-4 (1977), PA80-40159.
1490. Scholten, P. D., "The effects of ion-implanted gadolinium on the superconducting properties of thin niobium films", Thesis, The Florida State University, 1976, 81 pp., Order No. 77-08616.

1491. Scholten, P. D. and Moulton, W. G., "Effect of ion-implanted Gd on the superconducting properties of thin Nb-films", Phys. Rev. B., 15 (3), 1318-23 (1977), PA80-39986.
1492. Soeder, J. M. and Strizker, B., "Experiments on the formation of the A15-compounds Nb-Sn and Nb-Ge by ion implantation", J. Nucl. Mater., 72 (1-2), 270-4 (1978), EEA81-28791.
1493. Spiller, E., Haller, I., Feder, R., Baglin, J. E. and Hammer, W. N., "Graded-index AR surfaces made by ion implantation on plastic materials", Appl. Opt., 19 (17), 3022-6 (1980), PA84-6689.
1494. Vajda, P., Maury, F., Lucasson, A. and Lucasson, P., "Characterization of recovery stages in silver-implanted magnesium by subsequent electron irradiation", Proceedings of the International Conference on Ion Beam Modification of Materials, 1978, p. 1873-82.
1495. Yu, P. W., Grant, J. T., Park, Y. S. and Haas, T. W., "Zn-ion-implantation profiles in $CuInSe_2$ by Auger electron spectroscopy", J. Appl. Phys., 48 (1), 67-72 (1977), EEA80-20911.

VI. DEVICE APPLICATIONS

1. General

1496. Antipov, I., "Semiconductor devices having ion implanted and diffused regions", Patent USA 4155778, Publ. May 1979, CA91-48362.
1497. Bentini, G., "Ion implantation in semiconductors", Technol. Elettr., no. 12, 52-7 (1977), Italian, EEA81-38128.
1498. Bethge, K., "Applied research with heavy ions", Umsch. Wiss. & Tech., 77 (3), 84-5 (1977), German, EEA80-36243.
1499. Blood, P., "Ion implantation of semiconductor devices", Phys. Bull., 28 (9), 416-18 (1977), EEA81-10278.
1500. Bullard, R. L. and Hobbs, C. W., "Delineation of ion-implanted semiconductor devices for physical measurement", IBM Tech. Disclosure Bull., 21 (8), 3277 (1979), EEA82-38249.
1501. Cavaleru, A. and Dinca, G., "Some applications of ion bombardment in microelectronics", Electrochemical Society Spring Meeting (Extended abstracts), 1976, p. 526-8, EEA79-46107.
1502. Chernyaev, A. V. and Chernetskii, T. P., "Use of BF_2^+ implantation in the production of semiconductor devices", Sov. Microelectron., 8 (6), 377-80 (1979), EEA83-53666.
1503. Chu, C. K. and Johnson, J. E., "NTD silicon on high power devices", Neutron Transmutation Doping in Semiconductors Symposium, 1978, p. 53-63, EEA83-9721.
1504. De la Moneda, F., "Method to fabricate ion-implanted layers with abrupt edges to reduce the parasitic resistance of Schottky barrier FETs and bipolar transistors", Patent USA 4029522, Publ. June 1977, CA87-61587.

1505. Dearnaley, G., "Ion implantation of engineering components", Advances in Surface Coating Technology: An International Conference, Vol. 1., 1978, p. 111-13, PA82-83360.
1506. Dearnaley, G. and Hartley, N. E. W., "New uses of ion implanttation", Proceedings of the 4th Conference on the Scientific and Industrial Applications of Small Accelerators, 1976, p. 20-9, PA80-68989.
1507. DeNora, V. and Polata, B., "Reducing the defect density of an integrated circuit utilizing ion implantation", Patent USA 3976512, Publ. August 1976, CA85-136087.
1508. Ducamus, J., "Topology and implantation in LSI design", Onde Electr., 57 (4), 300-4 (1977), French, EEA80-24974.
1509. Duchynski, R. J., "Ion implantation for semiconductor devices", Solid State Technol., 20 (11), 53-8 (1977), EEA81-19429.
1510. Eisen, F. H. and Long, S. I., "Ion implanted GaAs I.C. process technology", Report AD-A078958, Rockwell International, Thousand Oaks, CA., 1979, 22 pp.
1511. Garcia, M., Piaguet, J., Roussin, J. C., Guernet, G. and Gilles, J., "Use of ion implantation for the creation of a P-type buried-layer structure in N-silicon: fabrication of complementary transistors", Vide, 171, 278-84 (1974), CA86-64387.
1512. Gibbons, J. F., "Ion implantation and proton-enhanced diffusion in semiconductors", J. Vac. Sci. & Technol., 12 (1), 475 (1975), PA79-2744.
1513. Goetzberger, A., "Masking process for producing periodic structures of small dimensions using ion implantation", Patent German 2454714, Publ. May 1975, CA85-86277.
1514. Gyulai, J., "Ion implantation research", Fiz. Sz., 26 (11), 423-7 (1976), Hungarian, EEA80-24864.
1515. Hashimoto, N. and Tamura, M., "Impurity doping [VLSI]", J. Inst. Electron. & Commun. Eng. Jpn., 62 (4), 410-15 (1979), Japanese, EEA83-4431.
1516. Janus, H. M., "Application of NTD silicon for power devices", Neutron Transmutation Doping in Semiconductors Symposium, 1978, p. 37-45, EEA83-9719.
1517. Karatsjuba, A. P., Kmita, T. G., Kruglow, I. I., Kurinny, V. I., Kurnosov, A. I., Ryzhikov, V. and Judin, V. V., "Improvements in semiconductor devices and methods for their manufacture", Patent UK 1435466, Publ. May 1976, CA85-136062.
1518. Kamoshida, M., "Ion implantation from the view point of micro fabrication for silicon devices", J. Jpn. Soc. Precis. Eng., 44 (2), 236-42 (1978), Japanese, EEA81-45937.
1519. Krylow, J., Bakowski, A. and Saczuk, K., "Examples of applications of ion implantation as a diffusion source in technologies of bipolar and MOS IC's", Proceedings of the International Conference on Ion Beam Modification of Materials, 1978, p. 239-66.

1520. Lee, D. H., Weller, K. P. and Thrower, W. F., "Planar process for making high frequency ion implanted passivated semiconductor devices and microwave integrated circuits", Patent USA 4030943, Publ. June 1977, CA87-61589.
1521. Lilen, H., "The dependence of power components of the future upon the application of neutron and/or electron bombardment techniques", Electron & Microelectron. Ind., No. 225, 22-5 (1976), French, EEA80-3457.
1522. Nicholas, K. H., "Applications of ion implantation in semiconductor devices", Acta Electron., 19 (2), 95-105 (1976), French, EEA79-32829.
1523. Nicholas, K. H., "Implantation damage in silicon devices", J. Phys. D., 10 (4), 393-407 (1977), EEA80-13457.
1524. Nicholas, K. H., "Device applications of ion beams", Proceedings of the International Conference on Ion Beam Modification of Materials, 1978, p. 915-40.
1525. Perkins, K. D., Walsh, P. S. and Ward, P. J., "A fully ion implanted integrated circuit process", Ist European Solid State Circuits Conference - ESSCIRC (Extended abstracts only), 1975, p. 113-14, EEA79-33152.
1526. Phillips, R. W., "The advantages of NTD silicon for high power semiconductor devices", Neutron Transmutation Doping in Semiconductors Symposium, 1978, p. 47-52, EEA83-9720.
1527. Ryssel, H., Haberger, K., Hoffmann, K., Prinke, G., Dumcke, R. and Sachs, A., "Simulation of doping processes", IEEE Trans. Electron Devices, ED-27 (8), 1484-92 (1980), EEA83-53646.
1528. Sansbury, J., "Applications of ion implantation in semiconductor processing", Solid State Technol., 19 (11), 31-7 (1976), EEA80-3465.
1529. Schwettmann, F. N., "Recent trends in device fabrication using ion implantation", J. Vac. Sci. & Technol., 13 (1), 390 (1976), PA79-58552.
1530. Signetics Corp., "Resistance with controlled temperature coefficient manufactured by ion implantation into a substrate", Patent French 2276688, Publ. Janluary 1976, CA85-135909.
1531. Sloan, B. J., Jr., "Fabricating integrated circuits utilizing ion implantation", Patent USA 3933528, Publ. January 1976, CA84-143802.
1532. Stephen, J., "Ion implantation", Electron, no. 85, 25, 27-8, 30 (1975), EEA79-12788.
1533. Stone, J. L. and Plunkett, J. C., "Recent advances in ion implantation - a state of the art review", Solid State Technol., 19 (6), 35-44 (1976), EEA79-41303.
1534. Wada, Y., Usui, H., Koyanagi, M. and Ashikawa, M., "Silicon device utilizing ion-implantation and selective oxidation", Patent USA 4016007, Publ. April 1977, CA86-181665.

2. p-n Junctions

1535. Akahane, R., Nitta, T. and Higuchi, H., "Quantitative study of implanted gold atoms in silicon p^+-n junctions by measurement of the thermally-stimulated current", Jpn. J. Appl. Phys., 18 (9), 1861-2 (1979), EEA83-674.

1536. Akimchenko, I. P., Vavilov, V. S., Krasnopevtsev, V. V., Milyutin, Y. V., Harsy, M. and Loi, C. K., "Properties of a p-n junction formed in n-type CdS by implantation of antimony ions", Sov. Phys.-Semicond., 9 (1), 19-21 (1975), PA79-5897.

1537. Anand, K. V., Sobhy, M. I. and El-Dhaher, A. H. G., "A study of the layer and junction properties of boron implantation in silicon", Int. J. Electron., 40 (2), 169-81 (1976), EEA79-16572.

1538. Armiento, C. A., Donnelly, J. P. and Groves, S. H., "P-N junction diodes in InP and $In_{1-x}Ga_xAs_yP_{1-y}$ fabricated by beryllium-ion implantation", Appl. Phys. Lett., 34 (3), 229-31 (1979), EEA82-26780.

1539. Armitage, S. A., "Radiation-enhanced diffusion of ion-implanted bismuth in cadmium sulphide", J. Phys. D, 8 (17), 2034-42 (1975), PA79-6409.

1540. Azikov, B. S., Brudnyi, V. N., Vyatkin, A. P., Krivov, M. A., Lupin, V. M. and Shirokov, L. L., "The influence of residual radiational defects on the characteristics of ion-doped gallium arsenide p-n junctions", Izv. VUZ Fiz., no. 7, 139-42 (1978), Russian, EEA81-45868.

1541. Balazs, A., Hermann, L. and Gyulai, J., "Determination of junction depth in implanted silicon by "pulled" anodization and capacitance-voltage measurements", Phys. Status Solidi A, 29 (2), K105-8 (1975).

1542. Bogardus, E. H., "Ion-implantation into As-diffused junctions in Si", J. Appl. Phys., 48 (5), 2093-4 (1977), EEA80-28777.

1543. Bogatyrev, V. A., Gavrilov, A. A., Kachurin, G. A., Pusep, Y. A. and Smirnov, L. S., "Electrical and photoelectric properties of p-n junctions in InSb formed by implantation of zinc ions followed by diffusion spreading", Sov. Phys.-Semicond., 12 (11), 1256-8 (1978), EEA82-34226.

1544. Bogatyrev, V. A., Gavrilov, A. A., Kachurin, G. A. and Smirnov, L. S., "Implanted p-n junctions in GaAs formed using laser pulse heating", Sov. Phys.-Semicond., 10 (7), 826-7 (1976), EEA80-9334.

1545. Bogatyrev, V. A., Kachurin, G. A. and Smirnov, L. S., "Properties in InSb p-n junctions fabricated by Zn implantation with subsequent drive-in diffusion", Proceedings of the International Conference on Ion Beam Modification of Materials, 1978, p. 277-86.

1546. Chu, M., Rahrenbruch, A. L., Bube, R. H. and Gibbons, J. F., "Photovoltaic properties of CdTe p-n junctions produced by ion implantation", J. Appl. Phys., 49 (1), 322-6 (1978), EEA81-24437.

1547. Devlin, W. J., Ip, K. T., Leta, D. P., Eastman, L. F., Morrison, G. H. and Comas, J., "P-N junction formation in n-InP by Be ion implantation", Symposium on Gallium Arsenide and Related Compounds 1978, p. 510-18, EEA83-14231.
1548. Fair, R. B. and Wivell, H. W., "Zener and avalanche breakdown in As-implanted low-voltage Si n-p junctions", IEEE Trans. Electron Devices, ED-23 (5), 512-18 (1976), EEA79-24764.
1549. Gavrilov, A. A., Kachurin, G. A., Bogomyakov, N. P., Kovalev, I. K. and Petrova, S. I., "Light emission of p-n junctions produced by implanted cadmium ions in gallium arsenide", Mikroelektronika (Akad. Nauk SSSR), 5 (2), 192-3 (1976), Russian, CA85-102983.
1550. Geipel, H., Tice, W. and Shasteen, R., "A correlation between the microstructure and electrical behaviour of implanted junctions", Electrochemical Society Fall Meeting, Extended Abstracts, 1976, p. 814-15, EEA80-36144.
1551. Helix, M. J., Vaidyanathan, K. V. and Streetman, B. G., "Properties of beryllium-implanted planar gallium arsenide p-n junctions", IEEE J. Solid-State Circuits, SC-13 (4), 426-9 (1978), CA89-172479.
1552. Kalinina, E. V., Prokof'eva, N. K., Suvorov, A. V., Kholuyanov, G. F. and Chelnokov, V. E., "Electrical properties of p-n junctions formed by ion implantation in n-type SiC", Sov. Phys.-Semicond., 12 (12), 1372-4 (1978), EEA83-9585.
1553. Kato, Y., Katayama, Y., Kobayashi, K. L. I. and Komatsubara, K. F., "P-N junction formation by Te^+ ion implantation into solution-grown $Pb_{1-x}Sn_xTe$", J. Appl. Phys., 46 (10), 4614-15 (1975), PA79-18894.
1554. Keresztes, P., Mohacsi, T. and Hegedus, A., "Theoretical model for calculation of breakdown voltage of implanted p-n junctions", Solid-State Electron., 19 (5), 415-18 (1976), CA85-12845.
1555. Kolodny, A. and Kidron, I., "Properties of ion-implanted junctions in mercury-cadmium-telluride", IEEE Trans. Electron Devices, ED-27 (1), 37-43 (1980).
1556. Lee, D. H., Weller, K. P., Ying, R. S. and Thrower, W. F., "Ion implantation process for fabricating high frequency avalanche devices", Patent USA 4064620, Publ. December 1977, CA88-82759.
1557. Lezheiko, L. V. and Lyubopytova, E. V., "Properties of p-n and n-n Si-β-SiC heterojunctions prepared by ion implantation", Sov. Phys.-Semicond., 10 (9), 1039-40 (1976), EEA80-20824.
1558. Lindner, M., "P-N junction formation by laser annealing of ion-implanted silicon", Phys. Status Solidi A, 57 (1), 263-7 (1980), EEA83-23771.
1559. MacIver, B. A. and Greenstein, E., "Damage effects in boron and BF_2 ion-implanted p^+-n junctions in silicon", J. Electrochem. Soc., 124 (2), 273-5 (1977), EEA80-13406.

1560. McMahon, R. A., Ahmed, H., Speight, J. D. and Dobson, R. M., "Scanning-electron-beam annealing of ion-implanted p-n junction diodes", Electron. Lett., 15 (14), 433-5 (1979), EEA82-38300.
1561. Mizutani, T. and Kurumada, K., "GaAs planar Gunn digital devices by sulphur-ion implantation", Electron. Lett., 11 (25-26), 638-9 (1975), EEA79-8864.
1562. Muller, J. C., Stuck, R. and Siffert, P., "TSC measurements on boron-implanted silicon junctions", Lattice Defects in Semiconductors, 1974, p. 513-20, Publ. 1975, EEA79-4644.
1563. Niu, H., Matsuda, T., Sadamatsu, H. and Takai, M., "Application of lateral photovoltaic effect to the measurement of the physical quantities of P-N junctions - sheet resistivity and junction conductance of N_2^+ implanted Si", Jap. J. Appl. Phys., 15 (4), 601-10 (1976), EEA79-24795.
1564. Park, Y. S., "Fabrication of ion implanted P-N junction devices", Report AD-D002428/1SL, Department of the Air Force, Washington, D. C., 1976, 21 pp.
1565. Rosinski, W. and Komicz, T., "Properties of Mn-ion implanted p-n junctions in n-type GaAs", Electron Technol., 7 (3-4), 69-72 (1974), EEA79-12873.
1566. Shedd, W. M., "Comparison of diffused and ion-implanted p-n junctions in radiation environments", Thesis, Northeastern University, 1977, 146 pp., Order No. 78-00296.
1567. Thom, R. D., Konkel, W. H. and Hoendervoogt, R. M., "Planar p-n junctions in InSb by Be ion implantation", IEEE Trans. Electron Devices, ED-24 (9), 1193-4 (1977), EEA81-1116.
1568. Tsai, M. Y., Morehead, F. F., Baglin, J. E. E. and Michel, A. E., "Shallow junctions by high-dose As implants in Si: experiments and modelling", J. Appl. Phys., 51 (6), 3230-5 (1980), EEA83-48699.
1569. Vavilov, V. S., Gukasyan, M. A., Guseva, M. I. and Konorova, E. A., "Conductivity of diamond doped by implantation of phosphorus ions", Sov. Phys.-Semicond., 9 (8), 962-4 (1975), EEA79-20246.
1570. Wada, Y. and Hashimoto, N., "Arsenic ion-implanted shallow junction", J. Electrochem. Soc., 127 (2), 461-6 (1980), EEA83-34675.
1571. Wu, C. P., Douglas, E. C., Mueller, C. W. and Williams, R., "Techniques for lapping and staining ion-implanted layers", J. Electrochem. Soc., 126 (11), 1982-8 (1979), EEA83-14225.
1572. Yu, P. W., Park, Y. S. and Grant, J. T., "Electroluminescence in Br-, Cl-, and Zn-implanted $CuInSe_2$ p-n junction diodes", Appl. Phys. Lett., 28 (4), 214-16 (1976), EEA79-25067.
1573. Zandweld, P., "Crystal damage and the properties of implanted p-n junctions in silicon", Solid-State Electron., 21 (5), 721-7 (1978), EEA81-33177.
1574. Zandveld, P., "Some properties of ion-implanted p-n junctions in silicon", Solid-State Electron., 19 (8), 659-67 (1976), PA79-72652.

3. Diodes

1575. Ajmera, P. K., "Electrical and optical studies of ion implanted infrared sensor diodes in epitaxially grown $Ga_xIn_{1-x}As$", Thesis, North Carolina State Univ., Raleigh, 1975, 128 pp., Order No. 76-886, CA84-143641.
1576. Ajmera, P. K. and Hauser, J. R., "Ion implanted photodiode detectors in epitaxial $Ga_xIn_{1-x}As$", Appl. Opt., 14 (12), 2905-10 (1975), EEA79-17071.
1577. Amith, A. and Mark, P., "Schottky barriers on ordered and disordered surfaces of GaAs (110)", J. Vac. Sci. & Technol., 15 (4), 1344-52 (1978), EEA82-732.
1578. Armstrong, B. M., Moore, R. A. and Wakefield, J., "Higher output power from Baritt diodes using ion implantation", Electron Lett., 15 (5), 153-5 (1979), EEA82-26782.
1579. Basilico, A., Fortino, A. G., Geipel, H. J. and Przekurat, K. D., "Ion-implanted buried finger photodiode", IBM Tech. Disclosure Bull., 21 (9), 3580 (1979), EEA82-38819.
1580. Berenz, J. J., Frank, F. B. and Hierl, T. L., "Ion-implanted p-n junction indium-phosphide IMPATT diodes", Electron Lett., 14 (21), 683-4 (1978), EEA82-4330.
1581. Bindell, J. B., Moller, W. M. and Labuda, E. F., "Ion-implanted low-barrier PtSi Schottky-barrier diodes", IEEE Trans. Electron Devices, ED-27 (2), 420-6 (1980), EEA83-39811.
1582. Bozler, C. O., Donnelly, J. P., Murphy, R. A., Laton, R. W., Sudbury, R. W. and Lindley, W. T., "High-efficiency ion-implanted lo-hi-lo GaAs IMPATT diodes", Appl. Phys. Lett., 29 (2), 123-5 (1976), EEA79-37528.
1583. Brack, K., Gaensslen, F. and Gansauge, P., "Ion-implanted Schottky barrier diodes", IBM Tech. Disclosure Bull., 19 (7), 2592 (1976), EEA80-32406.
1584. Chatterjee, P. K., Streetman, B. G., Keune, D. L. and Herzog, A. H., "Beryllium implanted gallium arsenide phosphide ($GaAs_{1-x}P_x$) light emitting diodes", Tech. Dig. - Int. Electron Devices Meet., 1975, p. 187-91, CA85-152815.
1585. Christou, A., Anand, Y. and Dietrich, H., "Low barrier height ion implanted X-band silicon Schottky barrier mixer diodes", Tech. Dig. - Int. Electron Devices Meet., 1977, p. 379-81, CA89-224905.
1586. Chu, W. K. and Sullivan, M. J., "Schottky barrier diode series resistance reduction using high energy phosphorus ion implantation", IBM Tech. Disclosure Bull., 22 (3), 988-9 (1979), EEA83-9771.
1587. Chu, W. K., Sullivan, M. J., Ku, S. M. and Shatzkes, M., "Low energy antimony implantation in silicon, II. Applications to Schottky barrier diode adjustment and to ion-implanted resistors", Proceedings of the International Conference on Ion Beam Modification of Materials, 1978, p. 193-210, Also: Radiat. Eff., 47 (1-4), 7-14 (1980), EEA83-44004.

1588. Donnelly, J. P., Armiento, C. A., Diadiuk, V. and Groves, S. H., "Planar guarded avalanche diodes in InP fabricated by ion implantation", Appl. Phys. Lett., 35 (1), 74-6 (1979), EEA82-47602.
1589. Donnelly, J. P. and Harman, T. C., "As^{+}-ion implanted lead telluride p-n junction photodiodes", Solid-State Electron., 18 (12), 1144-6 (1975), EEA79-5115.
1590. Eknoyan, O., Yang, E. S. and Sze, S. M., "Multilayered ion-implanted BARITT diodes with improved efficiency", Solid-State Electron., 20 (4), 291-5 (1977), CA86-198797.
1591. Hass, D., Heinrich, H., Jantsch, W., Lischka, K., Lopez-Otero, A., Palmetshofer, L. and Wagenhuber, M., "Comparison of PbTe diodes fabricated by epitaxial growth and by ion implantation of epitaxial layers", Crit. Rev. Solid State Sci., 5 (4), 547-53 (1975), EEA79-28391.
1592. Igras, E., Piotrowski, J. and Zimnoch-Higersberger, I., "Investigation of ion implanted graded GaP (CdHg)Te photodiodes", Electron Technol., 10 (4), 63-70 (1977), EEA81-19811.
1593. Ino, M., Ishibashi, T. and Ohmori, M., "C.W. oscillation with p^{+}-p-n^{+} silicon IMPATT diodes in 200 GHz and 300 GHz", Electron. Lett., 12 (6), 148-9 (1976), EEA79-20436.
1594. Ishiwara, H. and Furukawa, S., "Schottky barrier height control by using knock-on effect in ion implantation", Jpn. J. Appl. Phys., 16, Suppl. 1, 53-6 (1977), EEA80-36311.
1595. Kanamori, A., "Reduction in minority carrier storage effect by flourine ion implantation damage", IEEE Trans. Electron Devices, ED-25 (7), 722-8 (1978), EEA81-38169.
1596. Kolodny, A. and Kidron, I., "Properties of ion-implanted junctions in mercury-cadmium-telluride (n-p photodiode fabrication)", IEEE Trans. Electron Devices, ED-27 (1), 37-43 (1980), EEA83-40307.
1597. Kwor, Y. C. R., "Experimental and theoretical studies of ion-implanted silicon read IMPATT diode", Thesis, Cornell University, 1976, 144 pp., Order No. 76-15889.
1598. Law, H. D., Tomasetta, L. R. and Nakano, K., "Ion implanted InGaAsP avalanche photodiode", Appl. Phys. Lett., 33 (11), 920-2 (1978), EEA82-13102.
1599. Lee, D. H., Weller, K. P. and Thrower, W. F., "Passivated milli-meter-wave silicon IMPATT diodes fabricated by ion implantation", Proc. IEEE, 65 (2), 272-3 (1977), EEA80-9447.
1600. Lee, D. H., Weller, K. P. and Thrower, W. F., "Ion implanted planar-mesa IMPATT diodes for millimeter wavelengths", IEEE Trans. Electron Devices, ED-25 (6), 714-22 (1978), EEA81-38166.
1601. Li, S. S., Kim, J. S. and Wang, K. L., "Enhancement of effective barrier height in Ti-silicon Schottky diode using low-energy ion implantation", IEEE Trans. Electron Devices , ED-27 (7), 1310-12 (1980), EEA83-48759.

1602. Lindau, I., Chye, P. W., Garner, C. M., Pianetta, P., Su, C. Y. and Spicer, W. E., "New phenomena in Schottky barrier formation on III-V compounds", J. Vac. Sci. & Technol., 15 (4), 1332-9 (1978), EEA82-731.
1603. Lipinska, E., "Electroluminescent GaAs diodes made by means of zinc ion implantation", Arch. Electrotech., 26 (3), 639-42 (1977), Polish, EEA81-24447.
1604. Lui, S., Meyer, R. G. and Kwan, N., "An ion-implanted subsurface monolithic Zener diode", IEEE J. Solid-State Circuits, SC-14 (4), 782-4 (1979), EEA82-47130.
1605. MacIver, B. A., "Ion implanted Schottky barrier diode", Patent USA 4096622, Publ. June 1978, CA89-172598.
1606. MacIver, B. A. and Greenstein, E., "Forming diodes by amorphou implantations and concurrent annealing, monocrystalline reconversion and oxide passivation in (100) N-type silicon", Patent USA 4133704, Publ. January 1979, CA90-113975.
1607. MacIver, B. A. and Greenstein, E., "Low-dose phosphorus implantation for oxide-passivated diodes in (100) P-type silicon", Patent USA 4144100, Publ. March 1979, CA90-196554.
1608. Marine, J., Quillec, M. and Ravetto, M., "Zinc telluride electroluminescent diodes lubricated by ion implantation", Patent France 2281647, Publ. March 1976, CA86-49858.
1609. Marine, J., Schaub, B., Quillec, M., Ravetto, M. and Guernet, G., "Characterization of green electroluminescent diodes obtained by boron ion implantation in zinc telluride", Vide, 171, 312-20 (1974), CA86-64388.
1610. Milano, R. A., Helix, M. J., Windhorn, T. H., Streetman, B. G., Vaidyanathan, K. V. and Stillman, G. E., "Planar ion-implanted avalanche photodiodes in GaAs", Symposium on Gallium Arsenide and Related Compounds 1978, p. 411-19, EEA83-14817.
1611. Mizutani, T., Honda, T., Yamazaki, H. and Fugjimoto, M., "Gallium arsenide planar Gunn devices with sulfur-ion implanted n layers", Solid-State Electron., 20 (5), 443-5 (1977), CA87-32760.
1612. Murphy, R. A., Bozler, C. O., Donnelly, J. P., Laton, R. W., Lincoln, G. A., Sudbury, R. W., Lindley, W. T., Lowe, L. F. and Deane, M. I., "Ion-implanted lo-hi-lo annular GaAs IMPATT diodes", Proceedings of the 6th International Symposium on Gallium Arsenide and Related Compounds, Pt. II, 1976, p. 210-19, EEA81-6005.
1613. Ono, Y., Saito, K., Shiraki, Y. and Shimada, T., "$GaAs_{1-x}P_x$ light emitting diodes produced by Zn ion implantation. I. Dose dependence of properties of implanted diodes", Jpn. J. Appl. Phys., 14 (10), 1489-94 (1975), EEA79-1289.
1614. Plumb, R. G. and Carroll, J. E., "Thin silicon ion-implanted P-I-N photodiodes", IEE J. Solid-State and Electron Devices, 1 (3), 89-91 (1977), EEA80-25259.
1615. Schaumburg, H., "Application of the ion implantation technology to the production of varicap (varactor) diodes", Acta Electron., 19 (2), 147-9 (1976), EEA79-32995.

1616. Shiraki, Y., Shimada, T., Ono, Y. and Komatsubara, K. F., "$GaAs_{1-x}P_x$ light emitting diodes produced by Zn ion implantation. II. Photoluminescence of p-type layers formed by ion implantation and diffusion", Jpn. J. Appl. Phys., 14 (10), 1495-500 (1975), EEA79-1290.
1617. Shiraki, Y., Shimada, T., Ono, Y., Saito, K. and Komatsubara, K. F., "$GaAs_{1-x}P_x$ light-emitting diodes produced by Zn-ion implantation", Applications of Ion Beams to Materials, 1975, p. 86-90, Publ. 1976, EEA79-33473.
1618. Tissot, J. -L., Labrunie, G. and Marine, J., "Temperature behavior of ion-implanted ZnTe green LEDs", IEEE Trans. Electron Devices, ED-26 (8), 1202-5 (1979), EEA83-1138.
1619. Toyoda, N., Niikura, I., Shimura, Y., Hozuki, T., Sugibuchi, H., Mihara, M. and Hara. T., "Ion implanted GaAs varactor diodes: capacitance uniformity", Electron Lett., 14 (5), 152-4 (1978), EEA81-23939.
1620. Wilson, R. G. and Jamba, D. M., "Characteristics of Schottky barriers on n- and p-type implanted silicon", Electrochemical Society Fall Meeting, Extended Abstracts, 1976, p. 929-33, EEA80-36148.
1621. Wollschlaeger, K., "Study of the entry window of implanted silicon photodiodes", Zentralinst. Kernforsch., Rossendorf Dresden, ZfK-294, 226-7 (1975), German, CA86-82221.

4. Bipolar Transistors

1622. Akasaka, Y., Sakurai, H., Ono, N., Tsukamoto, K. and Hirao, T., "A high-speed bipolar process using full ion implantation", Mitsubishi Denki Giho, 53 (11), 786-8 (1979), Japanese, EEA83-29323.
1623. Akasaka, Y., Tsukamoto, K., Kawaguchi, M., Sato, H., Horie, K. and Komiya, H., "Application of diffusion from implanted polycrystalline silicon to bipolar transitors", Jap. J. Appl. Phys., 15, suppl. 15-1, 49-54 (1976), EEA80-6644.
1624. Akasaka, Y., Tsukamoto, K., Sakurai, T., Hirao, T., Horiba, Y., Kijima, K. and Nakata, H., "High speed bipolar process with full ion implantation and self-aligned contract structure (ECL RAM)", 1978 International Electron Devices Meeting, Technical Digest, p. 189-92, EEA83-5286.
1625. Alcorn, G. E. and Bergeron, D. L., "Method for making a double ion-implanted bipolar transistor having reduced corner injection", IBM Tech. Disclosure Bull., 22 (5), 1973-5 (1979), EEA83-19039.
1626. Anantha, N. G., Bhatia, H. S. and Walsh, J. L., "Method for fabrication vertical NPN and PNP structures utilizing ion implantation", Patent USA 4159915, Publ. July 1979.

1627. Ashburn, P., Bull, C., Nicholas, K. H. and Booker, G. R., "Effects of dislocations in silicon transistors with implanted bases", Solid-State Electron., 20 (9), 731-40 (1977), EEA80-41477.

1628. Assemat, J. L., Dumetz, A., Wattremez, J. M. and Lagorsse, J. M., "Effect of technological parameters on the characteristics of a 2.5 GHz hyperfrequency transistor combining diffusion and implantation", Vide, 171, 150-4 (1974), CA86-82476.

1629. Barile, C. A., Forneris, J. L., Kluge, H. and Seirmarco, J. A. "Ion implanting an emitter", IBM Tech. Disclosure Bull., 20 (11), 4351-2 (1978).

1630. Barile, C. A., Gaind, A. K. and Goth, G. R., "Ion implantation process for bipolar transitors", IBM Tech. Disclosure Bull., 20 (12), 5175-7 (1978).

1631. Barile, C. A., Goth, G. R. and Makris, J. S., "Ion-implanted base region formed after the emitter for bipolar transistors", IBM Tech. Disclosure Bull., 20 (5), 1736-7 (1977), EEA81-33335.

1632. Barile, C. A., Makris, J. S. and Peeters, R., "Ion implantation of emitter to eliminate arcing", IBM Tech. Disclosure Bull., 19 (10), 3727 (1977), EEA81-6089.

1633. Bartholomew, R. F., Putney, Z. C. and Stephens, G. B., "Ion implanted bipolar transistors for analog applications", Proceedings of Southeastcon '77 on Imaginative Engineering Thru Education and Experience, p. 211-14, EEA80-36442.

1634. Bergeron, D. L. and Stephens, G. B., "Complementary PNP transistor fabricated by ion implantation", IBM Tech. Disclosure Bull., 21 (4), 1550-1 (1978), EEA82-17603.

1635. Beyer, K. D., Das, G., Poponiak, M. R. and Yeh, T. H., "Reduction of emitter-collector leakages by ion implantation", International Electron Devices Meeting, Technical Digest, 1976, p. 42-4, EEA80-13504.

1636. Beyer, K. D., Poponiak, M. R. and Yeh, T. H., "Lower defect densities in implanted arsenic subcollector devices", IBM Tech. Disclosure Bull., 20 (3), 1003-4 (1977), EEA81-19423.

1637. Bhattacharyya, A., Selleck, J. E. and Wiedman, F. W., "Ion implantation of bipolar transistors having different characteristics", IBM Tech. Disclosure Bull., 18 (4), 1023-4 (1975), EEA79-967.

1638. Cho, K. H., "Buried collector process using nonselective ion implantation", Electrochemical Society Spring Meeting (Extended Abstracts), 1975, p. 325-6, EEA79-1083.

1639. Chu, W. K., Horng, C. T. and Schwenker, R. O., "Metal-silicide layer for the post emitter base implantation process", IBM Tech. Disclosure Bull., 21 (2), 580-1 (1978), EEA82-12643.

1640. Crippen, R. E., O'Brien, D., Rallapulli, K. and Verhofstadr, P. W. J., "Microprogram sequencer utilizing I^3L technology", 1976 International Solid-State Circuits Conference, Digest of Technical Papers, P. 98-9.

1641. Das, G., Gaind, A. K., Raheja, R. K. and Sugerman, A., "Fabrication of base region of a transistor using two ion implantation steps", IBM Tech. Disclosure Bull., 21 (2), 582-3 (1978), EEA82-12644.
1642. Doulin, J., Forneris, J. L., Kluge, H. C. and Regh, J., "Fabricating ion-implanted regions in semiconductors", IBM Tech. Disclosure Bull., 19 (9), 3407-8 (1977), EEA80-36427.
1643. Evans, S. A., Bartelt, J. L., Sloan, B. J., Jr. and Varnell, G. L., "Fabrication of integrated injection logic with electron-beam lithography and ion implantation", IEEE Trans. Electron Devices, ED-25 (4), 402-7 (1978), EEA81-28593.
1644. Forneris, J. L., Keenan, W. A. and Seirmarco, J. A., "Emitter ion implantation and diffusion process", IBM Tech. Disclosure Bull., 21 (4), 1434-5 (1978), EEA82-17405.
1645. Gegg, W. M., Saltich, J. L., Roop, R. M. and George, W. L., "Ion-implanted super-grain transistors", IEEE J. Solid-State Circuits, SC-11 (4), 485-91 (1976), EEA79-37541.
1646. Graul, J., Glasl, A. and Murrmann, H., "High-performance transistors with arsenic-implanted polysil emitters", IEEE J. Solid-State Circuits, SC-11 (4), 491-5 (1976), EEA79-37542.
1647. Graul, J., Glasl, A. and Murrmann, H., "Ion implanted bipolar high performance transistors with polysil emitter", 1975 International Electron Devices Meeting (Technical Digest), p. 450-4, EEA79-41535.
1648. Greiff, P., "Transistor design for high f_t at low currents", IEEE Trans. Electron Devices, ED-23 (3), 343-47 (1976).
1649. Hanson, J. W., Fordemwalt, J. N. and Huber, R. J., "Fabrication and performance of ion implanted I^2L devices", 1975 International Electron Devices Meeting (Technical Digest), p. 281-3, EEA79-46526.
1650. Hirao, T., Kijima, K. and Nakano, T., "Characteristics of Schottky TTL fabricated by oxide isolation utilizing ion-implantation", Electrochemical Society Spring Meeting, Extended Abstracts, 1977, p. 603-5, EEA81-19536.
1651. Hoffmeister, W., Schmitt, A., Schorer, G. and Schumacher, H., "Method of producing adjacent subcollector and subisolation areas", IBM Tech. Disclosure Bull., 22 (8A), 3253-4 (1980), EEA83-39745.
1652. Hu, S. M. and Schwenker, R. O., "Effects of ion implantation on substrate hardening and film stress reduction and their effect on the yield of bipolar transitors", J. Appl. Phys., 49 (6), 3259-65 (1978), EEA81-38123.
1653. Ko, W. and Schwenker, R. O., "An ion-implantation process for making bipolar devices", IBM Tech. Disclosure Bull., 19 (5), 1704-6 (1976), EEA80-17264.
1654. Koji, T., "Sequential anneal effect on bipolar transistor with phosphorous-implanted emitter", IEEE Trans. Electron Devices, ED-23 (9), 1103-4 (1976), EEA79-46246.

1655. Koji, T., Tseng, W. F. and Mayer, J. W., "Sidewall penetration of dislocations in ion-implanted bipolar transistors", Appl. Phys. Lett., 32 (11), 749-51 (1978), EEA81-38176.
1656. Lebesnerais, G. and Nuez, J. P., "Implanted transistor with lateral injection compensation", IBM Tech. Disclosure Bull., 20 (4), 1471 (1977), EEA81-23959.
1657. Monfret, A. and Bonis, M., "Ion implantation in the field of bipolar integrated circuits and components", Acta Electron., 19 (2), 129-39 (1976), French, EEA79-33100.
1658. Murakami, K., Ueda, M., Ohmori, M., Ohkura, I., Horiba, Y. and Nakano, T., "Double ion implanted DSAMOS-bipolar devices", 1977 International Electron Devices Meeting, Technical Digest, p. 186-9, EEA81-28583.
1659. Nagarajan, A. and Sharif, A., "Ion implantation of emitter region through thick oxides", IBM Tech. Disclosure Bull., 22 (11), 4888-9 (1980), EEA83-39753.
1660. Nakatsuka, M., Tanaka, K. and Kikkawa, T., "Effects of thin conductive film mask on ion implantation [bipolar ICs]", J. Electrochem. Soc., 125 (11), 1829-32 (1978), EEA82-22280.
1661. Ogureck, F. M. and Payne, R. S., "Fabrication of high-speed transistors by compensation implant near collector-base junction", Patent USA 4151009, Publ. April 1979, CA91-12688.
1662. Ohkura, M., Natsuaki, N., Tamura, M. and Tokuyama, T., "Single pulse laser annealing of a double-implanted layer", Jpn. J. Appl. Phys., 19 (2), L83-6 (1980), EEA83-23916.
1663. Onheiser, P., Zaruba, I., Rozsypal, A., Spetik, Z. and Chloupek, M., "Double ion implanted bipolar n-p-n transistors", Int. Conf. Ion Implantation Semicond., Proc., 1978, p. 574-85, CA90-96421.
1664. Parrillo, L. C. and Morris, B. L., "Deleterious effects of an oxidizing drive-in ambient on implanted arsenic emitters in (111) silicon", Appl. Phys. Lett., 35 (4), 345-7 (1979), EEA83-728.
1665. Payne, R. S., "A review of the application of ion-implantation to bipolar devices", Electrochemical Society Spring Meeting (Extended Abstracts), 1976, p. 517-20, EEA79-46260.
1666. Poponiak, M. R., Nagasaki, T. and Yeh, T. H., "Argon implantation gettering of bipolar devices", Electrochemical Society Fall Meeting (Extended Abstracts), 1975, p. 480, EEA79-16683.
1667. Poponiak, M. R., Nagasaki, T. and Yeh, T. H., "Argon implantation gettering of bipolar devices", J. Electrochem. Soc., 124 (11), 1802-5 (1977), EEA81-14620.
1668. Sander, W. B., Early, J. M. and Longe, T. A., "A 4096 x 1 (I^{3}I bipolar dynamic RAM", 1976 International Solid-State Circuits Conference, Digest of Technical Papers, p. 182-3.
1669. Sanders, I. R., "An ion implanted bipolar silicon integrated circuit process", Microelectron. & Reliab., 16 (1), 75-80 (1976), EEA80-21011.

1670. Seidel, T. E., Payne, R. S., Moline, R. A., Costello, W. R., Tsai, J. C. and Gardner, K. R., "Transistors with boron bases predeposited by ion implantation and annealed in various oxygen ambients", IEEE Trans. Electron Devices, ED-24 (6), 717-23 (1977), EEA80-24952.

1671. Tokuyama, T., Tamura, M., Yoshihiro, N., Ikeda, T. and Natsuaki, N., "High dose ion implantation into silicon (transistor production)", Oyo Buturi, 44 (6), 596-610 (1975), Japanese, PA79-17762.

1672. Tsuchimoto, T., Kumagaya, F. and Itoh, K., "Double implanted UHF power transistor with epitaxial overcoat contact to shallow emitter", Solid-State Electron., 19 (12), 1042-3 (1976), EEA80-6646.

1673. Tsukamoto, K., Akasaka, Y., Watari, Y., Kusano, Y., Hirose, Y. and Nakamura, G., "Arsenic-implanted emitter and its application to UHF power transitors", Proc. Conf. Solid State Devices, 1977, p. 187-92, CA89-98594.

1674. Wu, D. S., "Extraction of average doping density and junction depth in an ion-implanted deep-depletion transitor", IEEE Trans. Electron Devices, ED-27 (5), 995-7 (1980), EEA83-39878.

1675. Yagi, K., Tamura, M., Yanagi, Y., Inaniwa, K. and Tokuyama, T., "Low-frequency low-noise transistors fabricated by double ion implantation", Jpn. J. Appl. Phys., 19, suppl. 19-1, 61-4 (1979), EEA83-48783.

1676. Yamamoto, K., Matsushita, Y., Watanabe, M. and Takasu, S., "Induced defects by boron ion implantation and drive-in process", Electrochemical Society Spring Meeting (Extended Abstracts), 1976, p. 353-5, PA79-87928.

1677. Yuan, H. T., Doerbeck, F. H. and McLevige, W. V., "Ion implanted GaAs bipolar transistors", Electron. Lett., 16 (16), 637-8 (1980), EEA83-48772.

5. Field-Effect Transistors

A. General

1678. Anolick, E. S., "Improving oxide integrity by ion-implantation [FETs]", IBM Tech. Disclosure Bull., 20 (1), 287 (1977), EEA81-14762.

1679. Antipov, I. and Jadus, D. K., "Field-effect transistor with improved threshold value stability", Patent German 2545871, Publ. June 1976, CA85-86286.

1680. Clemens, J. T., Doklan, R. H. and Nolen, J. J., "N-channel Si-gate integrated circuit technology", Int. Electron Devices Meet., Tech. Dig., 1975, p. 299-302.

1681. De La Moneda, F. H. and Kotecha, H. N., "Fabrication of metal-gate depletion-mode load IGFETs with better defined implanted geometry", IBM Tech. Disclosure Bull., 21 (7), 2902-4 (1978), EEA82-22205.

1682. Dvurechensky, A. V., Gerasimenko, N. N., Romanov, S. I. and Smirnov, L. S., "High dose effects in ion implantation", Radiat. Eff., 30 (2), 69-71 (1976), PA79-87918.
1683. Edwards, J. R., "The ion-implanted depletion-mode IGFET", Electrochemical Society Spring Meeting, Extended Abstracts, 1975, p. 388-9, EEA79-4717.
1684. Hill, B. H., "Semiconductor device", Patent USA 3936860, Publ. February 1976, CA84-158844.
1685. Jacobus, L. A., Jr., "Insulated gate field effect transistor by implanted double counterdoping", Patent USA 4108686, Publ. August 1978, CA90-47498.
1686. Jaffe, J. M. and Tong, H. S., "Aluminum spiking prevention by ion implantation", Electrochemical Society Spring Meeting, Extended Abstracts, 1976, p. 253-5, EEA79-46324.
1687. Jambotkar, C. G., "Method for forming self-aligned field effect device by ion implantation and outdiffusion", Patent USA 4128439, Publ. December 1978.
1688. Joy, R. C., Magdo, I. E. and Phillips, A., "Process of making field effect transistor having improved threshold stability by ion implantation", Patent USA 4154626, Publ. May 1979.
1689. Luby, S., "The influence of oxygen ion implantation upon the lifetime of aluminum conductors for semiconductor devices", Electrotech. Cas., 31 (4), 317-24 (1980), Slovak, EEA83-39788.
1690. McKenna, C. M., "Electron flooding cone for ion implanting FET devices", IBM Tech. Disclosure Bull., 23 (4), 1403-4 (1980).
1691. McMahon, R. A. and Ahmed, H., "Direct device fabrication by selected area e-beam annealing", J. Vac. Sci. & Technol., 16 (6), 1840-2 (1979), EEA83-34768.
1692. Myers, D. R., Wilson, R. G. and Comas, J., "Considerations of ion channeling for semiconductor microstructure fabrication", J. Vac. Sci. & Technol., 16 (6), 1893-6 (1979), EEA83-34806.
1693. Nicholas, K. H., "Device applications of ion beams", Radiat. Eff., 47 (1-4), 121-9 (1980), EEA83-44013.
1694. Nuzillat, G. and Arnodo, C., "Adjusting the threshold voltage of field effect transistors", Patent Germany 2533460, Publ. February 1976, CA84-143821.
1695. Palkuti, L., Peng, J., Wells, C. and Teng, T. C., "Application of CW laser annealing in semiconductor device processing", Proceedings of the 30th Electronic Components Conference, 1980, p. 37-43, EEA83-34790.
1696. Rideout, V. L., "Process for providing self-aligned doping regions by ion implantation and lift off", Patent USA 4144101, Publ. March 1979.
1697. Sigmon, T. W. and Gibbons, J. F., "Applications of scanning continuous wave (CW) and pulsed laser in silicon technology", Proc. Soc. Photo-Opt. Instrum Eng., 198, 26-30 (1979), EEA83-48710.

1698. Smith, T. C., "Silicon production applications of ion implanttation", IEEE Trans. Nucl. Sci., NS-26 (1), pt. 2, 1677-82 (1979).
1699. Troutman, R. R., "Ion-implanted threshold tailoring for insulated gate field-effect transistors", IEEE Trans. Electron Devices, ED-24 (3), 182-92 (1977), EEA80-13510.
1700. Wada, Y. and Ashikawa, M., "Silicon planar devices using nitrogen ion implantation", Jpn. J. Appl. Phys., 15 (9), 1725-30 (1976), EEA79-46304.
1701. Wu, D. S., "Extraction of average doping density and junction depth in an ion-implanted deep-depletion transistor", IEEE Trans. Electron Devices, ED-27 (5), 995-7 (1980), EEA83-39878.

B. MOS

1702. Ahmed, H. and Charpentier, A., "Selective ion implantation to reduce power consumption in MOS integrated circuits", IEEE Trans. Electron Devices, ED-25 (5), 547-8 (1978), EEA81-28611.
1703. Alcorn, G. E., Kotecha, H. N. and Prosser, J. F., "Improved process for marking ion implanted junctions for MOSFET structures with high resistivity substrate", IBM Tech. Disclosure Bull., 21 (6), 2398-400 (1978), EEA82-22202.
1704. Anolick, E. S., "Improving oxide integrity by ion-implantation (FETS)", IBM Tech. Disclosure Bull., 20 (1), 287 (1977), EEA81-14762.
1705. Arnold, E., "Surface and bulk generation currents in ion-implanted MOS structures", 1977 International Electron Devices Meeting, Technical Digest, p. 529-31, EEA81-28376.
1706. Aymeloglu, S., Cheng, P. Y. and Zemel, J. N., "Freeze-out effects in ion-implanted MOSFETS", In: Semiconductor Silicon 1977, Huff, H. R. and Sirtl, E., (Eds.), Electrochemical Society, Princeton, N. J., p. 414-20 (1977), CA89-69543.
1707. Bandali, M. B. and Lo, T. C., "On modeling of the self-aligned field implanted MOS devices with narrow widths", 1975 International Electron Devices Meeting, Technical Digest, p. 573-6, EEA80-3522.
1708. Board, K., "The sensitivity of threshold voltage of ion-implanted MOSFET's to substrate doping and oxide thickness fluctuations", Int. J. Electron., 43 (2), 151-9 (1977), EEA80-36361.
1709. Bradbee, G. R., Oshelski, J. T. and Peterson, D. C., "Ion-implanted MOSFET", IBM Tech. Disclosure Bull., 20 (8), 3007-8 (1978), EEA82-4458.
1710. Brockman, H. E., "Improved yields for MOST's using ion implantation", J. Appl. Phys., 47 (4), 1716-18 (1976), CA84-188402.

1711. Brown, D. M., Connery, R. J. and Gray, P. V., "Doping profiles by MOSFET deep depletion C(V)", J. Electrochem. Soc., 122 (1), 121-27 (1975).

1712. Carrigan, D. G., Fu, H. S., Stephens, W. F., Tasch, A. F., Jr. and Cheek, T. F., Jr., "Low frequency noise characteristics of ion implanted 'buried channel' NMOS devices", IEEE Trans. Electron Devices, ED-24 (9), 1207 (1977), EEA81-1280.

1713. Cichomska, K., Wronski, W. and Oldziejewska, A., "Possibilities of using the $^{10}B(n,\alpha)^{7}Li$ reaction in structure studies of MOS [metal-oxide-semiconductor] integrated circuits", Elektronika, 18 (5), 209-11 (1977), Polish, CA87-176352.

1714. Demoulin, E. and Van de Wiele, F., "Ion implanted MOS transistors", Proceedings of the NATO Advanced Study Institute on Process and Device Modelling for Integrated Circuit Design, 1977, p. 617-76, EEA81-38237.

1715. Dennard, R. H. and Rideout, V. L., "N-channel MOSFETs having a common field/channel implant", IBM Tech. Disclosure Bull., 18 (4), 1289-91 (1975), EEA79-985.

1716. Dennard, R. H., Rideout, V. L., Yu, H. N. and Gaensslen, F. H., "Uses of ion implantation in advanced MOS field-effect transistors", Electrochemical Society Fall Meeting, Extended Abstracts, 1975, p. 326-9, EEA79-16889.

1717. Doucet, G., Van de Wiele, F. and Jespers, P., "Theoretical and experimental study of MOS transistors nonuniformly doped by silox technique", Solid-State Electron., 19 (3), 191-99 (1976).

1718. Duszak, M. and Jakubowski, A., "Influence of parameters of ion implantation process on shift of threshold voltage of MOS-transistor", Bull. Acad. Pol. Sci. Ser. Sci. Tech., 26 (10), 175-81 (1978), EEA82-34345.

1719. Edwards, J. R. and Marr, G., "Determination of ion-implanted profiles using the MOS capacitance-voltage technique", J. Vac. Sci. & Technol., 12 (6), 1373 (1976), PA79-36478.

1720. Fahrner, W. R. and Klausmann, E., "Characterization of MOS structures with buried layers", Phys. Status Solidi A, 50 (2), 433-44 (1978), EEA82-17360.

1721. Feltl, H., "Threshold voltage of m.o.s. transistors doped non-uniformly near the surface", IEE J. Solid-State & Electron Devices, 2 (6), 191-8 (1978), EEA82-4379.

1722. Fischer, W., Jacobs, E. P., Eisele, I. and Dorda, G., "Electron mobility in Si MOSFETs with an additional implanted channel", Solid-State Electron., 22 (3), 225-8 (1979), EEA82-22220.

1723. Gabler, L., Hoefflinger, B., Schneider, J. and Zimmer, G., "Extraction of implantation profiles from the differential body effect of ion-implanted m.o.s. transistors", Electron. Lett., 12 (10), 257-8 (1976), EEA79-28570.

1724. Geipel, H. J. and Shasteen, R. B., "Implanted source/drain junctions for polysilicon gate technologies", IBM J. Res. & Dev., 24 (3), 362-9 (1980), EEA83-39992.

1725. Goetzberger, A., Bartelink, D. J., McVittie, J. P. and Gibbons, J. F., "MOS measurement of oxygen recoils from As implantation into silicon dioxide", Appl. Phys. Lett., 29 (4), 259-61 (1976), PA79-84259.
1726. Goetzberger, A., Klausmann, E. and Schulz, M. J., "Dynamic properties of interface states in MOS-structures", Report AD-A059494, 1978, 35 pp., CA90-144919.
1727. Guernet, G., Garcia, M. and Bernard, J., "Ion implantation applied to MOS transistor technology", Acta Electron., 19 (2), 117-28 (1976), French, EEA79-33099.
1728. Horiuchi, S., "Threshold voltage shift of n-channel silicon-gate MOSFET's", IEEE Trans. Electron Devices, ED-22, 1038-43 (1975), CA84-37823.
1729. Itoh, K., Tsuchimoto, T. and Yudasaka, I., "Self-conducting MOS field effect semiconductor components", Patent Germany 2626362, Publ. January 1977, CA86-82633.
1730. Jhabvala, M., "Radiation hardened PMOS process with ion implanted threshold adjust", IEEE Trans. Nucl. Sci., NS-26 (1), 971-5 (1979), EEA82-22299.
1731. Jones, H. J. and Kotecha, H. N., "Partially ion-implanted MOSFET", IBM Tech. Disclosure Bull., 20 (7), 2746-7 (1977), EEA81-46077.
1732. Kennedy, D. P. and Kennedy, C. L., "Two dimensional analysis of ion implanted MOSFET devices", Proceedings of Southeastcon '80, p. 10-13, EEA83-34979.
1733. Kholomina, T. A., "Effect of irradiation with intermediate-energy ions on structure and electrophysical properties of silicon dioxide films in a MOS system", Sov. Phys. J., 21 (12), 1550-4 (1978), EEA83-4395.
1734. Klein, T., "Low voltage N-channel mosfet device", Patent USA 4104784, Publ. August 1978, CA89-189903.
1735. Koyanagi, M., Tamura, H., Miyao, M., Hashimoto, N. and Tokuyama, T., "Short-channel MOSFETs fabricated by self-aligned ion implantation and laser annealing", Appl. Phys. Lett., 35 (8), 621-3 (1979), EEA83-9817.
1736. Krimmel, E. F., "Self-adjusting preparation of doped regions in a semiconductor substrate by ion implantation", Patent Germany 2710266, Publ. September 1978, CA89-172609.
1737. Kubota, N., Shibata, T., Ohuchi, K., Hirabayashi, K. and Horiuchi, S., "New n-channel Si-gate technology for high packing density, high speed MOS LSI", Int. Electron Devices Meet., Tech. Dig., 1975, p. 295-98.
1738. Kudoh, O. and Kamoshida, M., "Drain current-voltage characteristics of ion-implanted MOS transistors", Trans. Inst. Electron & Commun. Eng. Jpn. Sect. E, E60 (11), 670-1 (1977), EEA81-14776.
1739. Ladbrooke, P. H., Huang, R. S. and Barnard, J. A., "Fast current pulse m.o.s. deep-depletion technique for profiling thin epitaxial and ion-implanted layers", IEE J. Solid-State & Electron Devices, 2 (4), 104-8 (1978), EEA81-45912.

1740. Landgraf-Dietz, D., "A simple method for the determination of parameters of ion implanted doping profiles by means of threshold voltage measurements of MOSFETs", Phys. Status Solidi A, 45 (2), K125-9 (1978), EEA81-24019.
1741. Lee, J. D., "Fabrication and evaluation of NMOS devices [using ion implantation techniques]", J. Korea Inst. Electron. Eng., 16 (4), 36-46 (1979), Korean, EEA83-24009.
1742. Lubberts, G. and Burkey, B. C., "Capacitance and doping profiles of ion-implanted, buried-channel MOSFETs", Solid-State Electron., 22 (1), 47-54 (1979), EEA82-22215.
1743. Lysenko, V. S. and Nazarov, A. N., "Formation and annealing of radiation damage in boron ion implanted MOS structures", Phys. Status Solidi A, 52 (1), 211-16 (1979), EEA82-30339.
1744. Lysenko, V. S. and Nazarov, A. N., "Radiation defects and impurity activation in MOS structures implanted with boron ions", Sov. Microelectron., 8 (6), 391-7 (1979), EEA83-53613.
1745. Lysenko, V. S., Nazarov, A. N., Kaschieva, S. B., Lokshin, M. P., Polishuk, Y. Y. and Ustilko, V. V., "Energy spectrum of radiation defects in MOS structures implanted with B ions of different energies", Ukr. Fiz. Zh., 23 (8), 1379-82 (1978), Russian, EEA81-45896.
1746. Lysenko, V. S., Sachenko, A. V., Dubchak, A. P., Nazarov, A. N., Altamanenko, V. T., Lokshin, M. P. and Polishuk, Y. Y., "Properties of MOS structures after B^+ ion bombardment", Phys. Status Solidi A, 38 (1), 131-8 (1976), EEA80-9374.
1747. Lysenko, V. S., Sachenko, A. V., Nazarov, A. N. and Kaschieva, S. B., "Effect of implanted B^+ ions on the voltage-capacitance characteristics of MOS-structures", Bulg. J. Phys., 5 (1), 57-62 (1978), Russian, EEA81-42289.
1748. Meiling, G. S. and Cauge, T. P., "Metal oxide semiconductor structures using ion implantation", Patent USA 4001048, Publ. January 1977, CA86-64519.
1749. Merckel, G. "Ion implanted MOS transistors-depletion mode devices", Proceedings of the NATO Advanced Study Institute on Process and Device Modelling for Integrated Circuit Design", 1977, p. 677-88, EEA81-38238.
1750. Miyake, M. and Harada, H., "Fabrication of MOS LSI silicon substrates by ion implantation", Trans. Inst. Electron. and Commun. Eng. Jpn. Sect. E, E63 (7), 558 (1980), EEA84-1481.
1751. Miyao, M., Koyanagi, M., Tamura, H., Hashimoto, N. and Tokuyama, T., "Short channel MOSFETs fabricated by self-aligned ion implantation and laser annealing", Jpn. J. Appl. Phys., 19, suppl. 19-1, 129-32 (1979), EEA83-48815.
1752. Murase, K. and Harada, H., "High dose As ion implantation for MOSFET fabrication", Trans. Inst. Electron. and Commun. Eng. Jpn. Sect. E, E62 (10), 709-10 (1979), EEA83-19060.
1753. Murel, A. V., Nikandrov, A. I., Shitova, E. V. and Zorin, E. I., "Relaxation of nonequilibrium capacitance in MOS structures irradiated with argon ions", Mikroelektronika, 6 (3), 276-9 (1977), Russian, EEA81-33239.

1754. Nassibian, A. G., Browne, V. A. and Perkins, K. D., "Generation lifetime investigation of ion-damage gettered silicon using MOS structure", J. Appl. Phys., 47 (3), 992-6 (1976), PA79-54457.
1755. Nassibian, A. G. and Golja, B., "Lifetime in ion implant damage gettered silicon by MOS voltage ramping", IEEE Trans. Electron Devices, ED-26 (3), 245-6 (1979), EEA83-34796.
1756. Nassibian, A. G. and Golja, B., "Investigation of Ar ion implant gettering of gold in silicon by MOS and Rutherford backscattering techniques", IEE Proc. I, 127 (1), 29-36 (1980), EEA83-18862.
1757. Niggebrugge, U., Tsironis, C., Filensky, W., Balk, P. and Beneking, H., "Implanted Si m.o.s.f.e.t.s for high-speed application", Electron. Lett., 14 (21), 690-1 (1978), EEA82-864.
1758. Nihira, H., Konaka, M., Iwai, H. and Hishi, Y., "Anomalous drain current in N-MOSFET's and its suppression by deep ion implantation (for VLSI)", 1978 International Electron Devices Meeting, Technical Digest, p. 487-91, EEA83-4562.
1759. Ohkura, I., Ohmori, M., Shimotori, K., Nakano, T., Hayashi, Y. and Tarui, Y., "Fully ion implanted DSA MOS IC", Jpn. J. Appl. Phys., 16, suppl. 1, 167-71 (1977), EEA80-36454.
1760. Okabe, T., Ochi, S., Kurono, H. and Kagami, J., "Short channel, low noise UHF MOS-FET's utilizing molybdenum-gate masked ion-implantation", Jap. J. Appl. Phys., 15, suppl. 15-1, 201-6 (1976), EEA80-9482.
1761. Oleszek, G. M., Perner, F. A., Spencer, O. S. and Wang, P. P., "Ion-implanted high voltage MOSFET device", IBM Tech. Disclosure Bull., 21 (2), 681-2 (1978), EEA82-12678.
1762. Ott, G., "Self-conducting MOS-field effect transistor", Patent Germany 2638867, Publ. March 1978, CA88-145176.
1763. Petrov, I. N., Glukharov, T. N., Georgiev, V. K. and Nencheva, G. P., "Theoretical analysis of the threshold voltage in ion-implanted p-channel MOS transistors", Bulg. J. Phys., 4 (3), 280-90 (1978), EEA81-19365.
1764. Rousseau, P. and Bernard, J., "Application of ion implantation to the fabrication of complex integrated circuits comprised of MOS transistors", Vide, 171, 85-91 (1974), CA86-82474.
1765. Runge, H., "Threshold voltage shift of MOS transistors by ion implantation of boron, aluminum, gallium, phosphorus, and arsenic", In: Ion Implantation Semicond. Sci. Technol. Proc. Int. Conf. 4th, Namba, S., (Ed.), p. 703-8 (1975), CA84-11406.
1766. Runge, H., "Device geometry dependent effects of deep levels on C(U) measurements of ion-implanted MOS diodes", Phys. Status Solidi A, 39, 141-6 (1977), CA86-99825.
1767. Saito, K., "Ion-implanted self-aligned Mo gate MOS-FET", Trans. Inst. Electron. & Commun. Eng. Jpn. Sect. E, E60 (9), 503-4 (1977), EEA81-10362.

1768. Schemmert, W., "Charge transport in ion implanted buried channel MOS transistors", Arch. Elektron. & Uebertragungstech., 33 (1), 23-31 (1979), German, EEA82-22176.
1769. Schemmert, W., Gabler, L. and Hoefflinger, B., "Conductance of ion-implanted buried-channel MOS transistors", IEEE Trans. Electron Devices, ED-23 (12), 1313-19 (1976), EEA80-6662.
1770. Serzec, B., Lagorsse, J. M., Assemat, J. L. and Lecomte, J., "Self-aligned transistors. Correlation of the electrical and implantation parameters. Control of the threshold voltage of MOS transistors", Vide, 171, 69-74 (1974), French, CA86-99440.
1771. Shah, P., Shah, R. and Crosthwait, L., "MOS device and material characterization of laser annealed implanted polysilicon", 1979 International Electron Devices Meeting, Technical Digest, p. 216-19.
1772. Shibata, T. and Harada, N., "MOSFET devices by ion-implantation", Patent USA 4140547, Publ. February 1979, CA90-179057.
1773. Shimotori, K., Anami, K., Nagayama, Y., Ohkura, I., Ohmori, M., Nakano, T., Hayashi, Y. and Tarui, Y., "Fully ion implanted 4096-bit high speed DSA MOS RAM", 1977 International Solid-State Circuits Conference, Digest of Technical Papers, p. 76-7, EEA80-36484.
1774. Sigmon, T. W., "Stabilization of MOS structures by boron ion implantation", Proc. IEEE, 63 (11), 1619-20 (1975), PA79-2171.
1775. Spadea, G., "Ion implantation method", Patent Germany 2844162, Publ. April 1979, CA90-214173.
1776. Teng, T. C., Merritt, J. D., Velez, J., Peng, J. and Palkuti, L., "Short-channel polysilicon-gate m.o.s.f.e.t.s. fabricated by c.w. argon laser annealing of arsenic-implanted source and drain", Electron. Lett., 16 (12), 477-80 (1980), EEA83-44090.
1777. Topich, J. A., "Temperature sensitivity of ion-implanted MOS capacitors", Appl. Phys. Lett., 34 (11), 787-9 (1979), EEA82-38137.
1778. Tsironis, C. and Niggebrugge, U., "High frequency performance of implanted Si-MOSFETs", 1979 IEEE MIT-S International Microwave Symposium Digest, p. 396-8, EEA82-38366.
1779. Van de Wiele, F., Demoulin, E., Sinon, R. and Gilles, C., "Models for implanted MOS transistors", Modelling Semiconductor Devices Symposium, 1977, p. 325-38, EEA81-23930.
1780. Wada, Y. and Nishimatsu, S., "Application of high current arsenic ion implantation to dynamic MOS memory LSI's", Jpn. J. of Appl. Phys., 18, suppl. 1, 247-54 (1979).
1781. Wada, Y., Nishimatsu, S. and Sato, K., "Evaluation of arsenic implanted layers by means of MOS memory characteristics", Solid-State Electron., 21 (3), 513-18 (1978), CA89-35235.
1782. Wang, K. L., "Relationships between residual defects and excess noise in ion-implanted MOSFETs", International Electron Devices Meeting, Technical Digest, 1976, p. 335-9, EEA80-36390.

1783. Wang, K. L., "Defect spatial distributions in annealed ion-implanted silicon measured by a transient capacitance technique [MOSFET]", Appl. Phys. Lett., 29 (11), 700-2 (1976), EEA80-6655.
1784. Wang, K. L., "Measurements of residual defects and 1/f noise in ion-implanted p-channel MOSFET's", IEEE Trans. Electron Devices, ED-25 (4), 478-84 (1978), EEA81-28516.
1785. Wang, P. P., "Double boron implant short-channel MOSFET", IEEE Trans. Electron Devices, ED-24 (3), 196-204 (1977), EEA80-13511.
1786. Wang, P. P. and Spencer, O. S., "Threshold voltage characteristics of double-boron-implanted enhancement-made MOSFETs [metal-oxide-semiconductor field-effect transistor]", IBM J. Res. Dev., 19 (6), 530-8 (1975), CA84-68531.
1787. Wang, S. T., Royce, B. S. H. and Russel, T. J., "The effect of ion-implantation on oxide charge storage in MOS devices", IEEE Trans. Nucl. Sci., NS-22, 2168-73 (1975), CA84-114817.
1788. Yoshida, I., Kubo, M. and Ochi, S., "High power MOSFET with a vertical drain electrode and meshed gate structure", Int. Electron Devices Meet., Tech. Dig., 1975, p. 159-62.
1789. Zimmer, G., "Laser annealing of MOS-transistor channel implantations", Electron. Lett., 15 (6), 184-6 (1979), EEA82-22180.
1790. Zohta, Y., "C-V and G-V characteristics of ion-implanted MOS structures depending upon the geometrical structure of the implanted region", J. Appl. Phys., 48 (4), 1615-20 (1977), EEA80-24820.

C. GaAs

1791. Ch'en, D. R., "Ion implanted GaAs FET devices and circuits", Eleventh Annual Asilomar Conference on Circuits Systems and Computers, 1977, p. 134-8, EEA81-38232.
1792. Ch'en, D. R. and Eisen, F. H., "Ion implanted high frequency high speed GaAs integrated circuits technology", J. Vac. Sci. & Technol., 16 (6), 2054-62 (1979), EEA83-40007.
1793. Doerbeck, F. H., Macksey, H. M., Brehm, G. E. and Frensley, W. R., "Ion-implanted GaAs X-band power FET", Electron. Lett., 15 (18), 576-8 (1979), EEA82-47174.
1794. Eden, R. C. and Welch, B. M., "Low power depletion mode ion-implanted GaAs FET integrated circuits", IEEE Trans. Electron Devices, ED-24 (9), 1209-10 (1977), EEA81-6101.
1795. Higgins, J. A., Eisen, F., Welch, B., Robinson, G. and Hill, W., "GaAs FETs fabricated by selenium ion implantation", Proceedings of the 6th International Symposium on Gallium Arsenide and related compounds, Pt. II, 1976, p. 236-44, EEA81-6057.

1796. Higgins, J. A., Kuvas, R. L. and Ch'en, D. R., "Modeling, fabrication and performance of ion implanted low-noise GaAs FETs", 1977 International Electron Devices Meeting, Technical Digest, p. 506-9, EEA81-28552.

1797. Higgins, J. A., Kuvas, R. L., Eisen, F. H. and Ch'en, D. R., "Low noise GaAs FETs prepared by ion implantation", IEEE Trans. Electron Devices, ED-25 (6), 387-96 (1978), EEA81-38215.

1798. Higgins, J. A., Kuvas, R. L., Hill, W. A., Robinson, G. D. and Ch'en, D. R., "Improved noise performance of ion implanted gallium arsenide FETs", Proc. - Bienn. Cornell Electr. Eng. Conf., 1977, p. 337-46, CA89-98599.

1799. Higgins, J. A., Welch, B. M., Eisen, F. H. and Robinson, G. D., "Performance of sulphur-ion-implanted GaAs f.e.t.s.", Electron. Lett., 12 (1), 17-18 (1976), EEA79-8927.

1800. Higgins, J. A., Welch, B. M., Eisen, F. H. and Robinson, G. D., "Performance of selenium-ion-implanted gallium arsenide F.E.T.s", Electron. Lett., 12 (18), 462-4 (1976), CA85-170574.

1801. Immorlica, A. A., Jr., Higgins, J. A., Hill, W., Robinson, G. and Kuvas, R. L., "Ion implanted GaAs FETs", 1978 International Electron Devices Meeting, Technical Digest, p. 368-72, EEA83-4552.

1802. Ishii, Y., Nojima, S., Ida, M. and Kurumada, K., "A new approach to fabrication of GaAs JFETs [using Be ion implantation]", Solid-State Electron., 23 (3), 269-73 (1980), EEA83-29269.

1803. Kellner, W., Kniepkamp, H., Ristow, D. and Boroffka, H., "Microwave field-effect transistors from sulfur-implanted gallium arsenide", Tech. Dig. - Int. Electron Devices Meet., 1975, p. 238-42, CA85-152816.

1804. Kung, J. K., Malbon, R. M. and Lee, D. H., "GaAs FETs with silicon-implanted channels", Electron. Lett., 13 (7), 187-8 (1977), EEA80-17238.

1805. Kuvas, R. L., Higgins, J. A. and Ch'en, D. R., "Ion implanted gallium arsenide power field effect transistors", Report AD-A063397, Sci. Cent., Rockwell Int., Thousand Oaks, CA, 1978, 71 pp., CA91-82121.

1806. Ladd, G. O., Krumm, C. F., Gray, W. W. and Keithley, G. W., "Medium power gallium arsenide FET by ion implantation", Proc. - Bienn. Cornell Electr. Eng. Conf., 1977, p. 289-95, CA89--98598.

1807. Ladd, G. O., Jr., Lundgren, R. E., Cleary, F. C. and Hirsch, N., "Low-noise, 17-GHz GaAs FET with ion-implanted channel", 1976 IEEE International Solid-State Circuits Conference, Digest of Technical Papers, 1976, p. 169, EEA79-46318.

1808. Li, C. T., Chen, P. T. and Wang, P. H., "A K-band ion implanted GaAs FET", Proceedings of the 8th European Microwave Conference, 1978, p. 396-400, EEA82-17541.

1809. Morgan, D. V., "The characterization and application of ion-induced damage in gallium arsenide devices", Radiat. Phys. & Chem., 15 (5), 627-36 (1980), EEA83-44017.

1810. Ohmachi, Y. and Nishioka, T., "Ion-implanted n-channel InP IGFET and its low frequency characteristics", Jpn. J. Appl. Phys., 19 (7), 1425-6 (1980), EEA83-48811.
1811. Sleger, K. J. and Dietrich, H. B., "GaAs FET device fabrication and ion implantation technology (microwave III-V compound semiconductors)", Report AD-A070783/6SL, Naval Research Lab, Washington, DC, 1979, 29 pp.
1812. Stoneham, E., Tan, T. S. and Gladstone, J., "Fully ion implanted GaAs power FETs", 1977 International Electron Devices Meeting, Technical Digest, p. 330-3, EEA81-28542.
1813. Vodicka, V. W. and Zuleeg, R., "Ion implanted GaAs enhancement mode JFET's", 1975 International Electron Devices Meeting, Technical Digest, p. 625-8, EEA80-6675.
1814. Wiley, J. D., "C-V profiling of GaAs FET films", IEEE Trans. Electron Devices, ED-25 (11), 1317-24 (1978), EEA82-22187.
1815. Zuleeg, R. and Lehovec, K., "Neutron degradation of ion-implanted and uniformly doped enhancement mode GaAs JFETs", IEEE Trans. Nucl. Sci., NS-25 (6), 1444-9 (1978), EEA82-12675.

D. SOS

1816. Alestig, G., Holmen, G., Martenson, M. and Peterstrom, S., "Electrical measurements of boron implanted silicon on sapphire and bulk silicon", Proceedings of the International Conference on Ion Beam Modification of Materials, 1978, p. 211-24.
1817. Inoue, T. and Yoshii, T., "Crystalline disorder reduction and defect-type change in silicon on sapphire films by silicon implantation and subsequent thermal annealing", Appl. Phys. Lett., 36 (1), 64-6 (1980), EEA83-23812.
1818. Ipri, A. C. and Flatley, D. W., "Radiation tolerant silicon gate CMOS/SOS using ion implantation", IEEE Trans. Electron Devices, ED-23 (9), 1110-12 (1976), EEA79-46429.
1819. Jamba, D. M., Wilson, R. G. and Harari, E., "Characterization of insulating regions created in silicon-on-sapphire by high dose ion implantation", Electrochemical Society Fall Meeting, Extended Abstracts, 1976, p. 500, EEA80-36469.
1820. Mizokami, H., Ino, M. and Hashimoto, T., "Ion implantation and HCl treatment simplify SOS/CMOS processing and slash leakage current", JEE, no. 103, p. 26-30 (1975), PA79-2745.
1821. Peterstrom, S. and Holmen, G., "Ion implanted devices in silicon on sapphire", Phys. Scr., 22 (3), 308-13 (1980), EEA83-44001.
1822. Repace, J. L., "Radiation effects in ion-implanted SOS capacitors with negative charge", IEEE Trans. Nucl. Sci., NS-25 (6), 1450-3 (1978), EEA82-12756.

1823. Splittgerber, H. and Takacs, D., "Doping of ESFI SOS transistors with ion-implantation", Electrochemical Society Spring Meeting, Extended Abstracts, 1976, p. 358-60, EEA79-46516.

E. CMOS

1824. Aitken, A., Poulsen, R. G., MacArthur, A. T. P. and White, J. J., "A fully plasma etched ion implanted CMOS process", International Electron Devices Meeting, Technical Digest, 1976, p. 209-13.
1825. Izumi, K., Doken, M. and Ariyoshi, H., "C.M.O.S. devices fabricated on buried silicon dioxide layers formed by oxygen implantation into silicon", Electron. Lett., 14 (18), 593-4 (1978), CA89-224884.
1826. Izumi, K., Doken, M. and Ariyoshi, H., "High speed CMOS IC using buried SiO_2 layers formed by ion implantation", Jpn. J. Appl. Phys., 19, suppl. 19-1, 151-4 (1979), EEA83-53813.
1827. Izumi, K., Doken, M., Yamamoto, E. and Irita, Y., "Completely isolated CMOS ICs fabricated by oxygen ion implantation technique called 'SIMOX'", Trans. Inst. Electron. and Commun. Eng. Jpn. Sec. E, E62 (12), 921-2 (1979), EEA83-29361.
1828. Motamedi, M. E., Yee, T. K. and Steckl, A. J., "Design and evaluation of ion-implanted CMOS structures", IEEE Trans. Electron Devices, ED-27 (3), 578-83 (1980), EEA83-39977.
1829. Parrish, W. J., "An ion-implanted CMOS amplifier for high performance active filters", Thesis, University of California, Santa Barbara, 1976, 200 pp., Order No. 77-04877.

F. MIS

1830. Brown, W. D., "Metal-nitride-oxide-semiconductor transistor characterization using implanted MNOS capacitors", Thin Solid Films, 59 (1), 125-9 (1979), EEA82-42531.
1831. Crowder, B. L., Hunter, W. R. and Ormond, D. W., "Method of manufacturing semiconductor devices in which oxide regions are formed by an oxidation mask disposed directly on a substrate damaged by ion implantation", Patent USA 4098618, Publ. July 1978.
1832. Gosney, W. M. and Buehler, M. G., "Metal-insulator-semiconductor structures having reduced junction capacitance", Patent USA 3852119, Publ. December 1974, CA82-79553.
1833. Ito, T., Hijiya, S., Nishi, H., Shinoda, M. and Furuya, T., "Interfacial doping by recoil implantation for nonvolatile memories", Jpn. J. Appl. Phys., 17, suppl. 17-1, 201-6 (1977), EEA81-46084.

1834. Lysenko, V. S. and Nazarov, A. N., "Effect of laser beam irradiation on the electrophysical properties of the interface in boron ion implanted MIS structures", Phys. Status Solidi A, 49 (1), 405-11 (1978), CA89-207933.
1835. Lysenko, V. S., Nazarov, A. N., Lokshin, M. M. and Kaschieva, S. B., "Influence of laser radiation on electrical properties of insulator-semiconductor interfaces in metal-insulator-semiconductor structures doped by B^+ ion implantation", Sov. Phys.-Semicond., 11 (11), 1327-8 (1977), EEA81-38097.
1836. Nakamura, K. and Kamoshida, M., "Electrical characteristics of $^{11}B^+$-implanted p-channel MNOS transistors", Solid-State Electron., 18 (12), 1085-8 (1975), EEA79-4711.
1837. Sefick, S. A. and Jones, R. K., "Narrow channel field effect transistor by masking and ion-implantation", Patent USA 4145233, Publ. March 1979, CA90-196557.
1838. Turley, A. P., Molouski, R. M., Reid, P. R. and McCann, D. H., "Fabrication of p-channel silicon gate LSI devices with Si_3N_4/SiO_2 gate dielectric", J. Electrochem. Soc., 123 (1), 117-20 (1976).
1839. Uchida, Y., Endo, N., Saito, S. and Nishi, Y., "Avalanche-tunnel injection in MNOS transistor", IEEE Trans. Electron Devices, ED-24 (6), 688-93 (1977), EEA80-24951.

G. MESFET

1840. Buczkowski, A., Kramkowska, M., Ohly, T. and Szreter, M., "Application of ion implantation to FET production", Pr. Nauk. Inst. Technol. Elektron. Politech. Wroclaw, 21, 83-7 (1978), Polish, CA90-161038.
1841. de Santis, P., "Extension of existing models to ion-implanted MESFET's", IEEE Trans. Microwave Theory and Tech., MTT-28 (6), 638-47 (1980), EEA83-48803.
1842. Feltl, H., "Fine-geometry Schottky-gate FET in ion implantation technology for LSI", Siemens Forsch. Entwicklungsber., 5 (3), 136-8 (1976), CA85-152732.
1843. Kellner, W., Kniepkamp, H., Ristow, D., Heinzle, M. and Boroffka, H., "Microwave field-effect transitors from sulphur-implanted GaAs", Solid-State Electron., 20 (5), 459-62 (1977), EEA80-24962.
1844. Kitahara, K. Nakai, K., Shibatomi, A. and Ohkawa, S., "Substrate bias effect on ion-implanted GaAs MESFET's", Jpn. J. Appl. Phys., 19 (7), 1369-71 (1980), EEA83-48812.
1845. Nozaki, T. and Ohata, K., "Sub-micron gate GaAs MESFET's with ion-implanted channels", Jpn. J. Appl. Phys., 16, suppl. 1, 111-14 (1977), EEA80-36366.
1846. Ohata, K., Nozaki, T. and Kawamura, N., "Improved noise performance of GaAs MESFET's with selectively ion-implanted n^+ source regions", IEEE Trans. Electron Devices, ED-24 (8), 1129-303 (1977), EEA80-36360.

1847. Taylor, G. W., Darley, H. M., Frye, R. C. and Chatterjee, P. K., "Device model for an ion-implanted MESFET", IEEE Trans. Electron Devices, ED-26 (3), 172-82 (1979), EEA83-34941.
1848. Vodicka, V. W., Zuleeg, R. and Cernjavic, D., "Ion-implanted GaAs Schottky-barrier FETs", Monitor, 37 (10), 311-14 (1976), EEA80-24959.

H. Others

1849. Akasaka, Y., Horie, K., Mitarai, G., Hirose, Y., Nomura, K. and Nishiumi, H., "Antimony (+)-implanted buried layer beneath thick oxide applied for vertical FET", In: Ion Implantation Semicond., 5th, Chernow, F., Borders, J. A. and Brice, D. K., (Eds.), p. 535-42 (1977), CA89-83665.
1850. Baudot, W. and Sansen, W., "DC modelling of ion-implanted JFETs", Symposium on Modelling Semiconductor Devices, 1977, p. 363-73, EEA81-23923.
1851. Fuoss, D. and Verma, K., "A fully implanted V-groove power MOSFET", 1978 International Electron Devices Meeting, Technical Digest, p. 657-60, EEA83-4568.
1852. Gleason, K. R., Dietrich, H. B., Bark, M. L. and Henry, R. L., "Enhancement-mode ion-implanted InP FETs", Electron Lett., 14 (19), 643-4 (1978), EEA81-49787.
1853. Gleason, K. R., Dietrich, H. B., Henry, R. L., Cohen, E. D. and Bark, M. L., "Ion-implanted n-channel InP metal semiconductor field-effect transistor", Appl. Phys. Lett., 32 (9), 578-81 (1978), EEA81-38202.
1854. Kamins, T. I. and Rose, P. H., "Electron-beam annealing of ion-implantation damage in integrated-circuit devices", J. Appl. Phys., 50 (3), pt. 1, 1308-11 (1979), EEA82-38252.
1855. Mizutani, T., Ishida, S., Kurumada, K. and Ohmori, M., "Selectively ion implanted Gunn-f.e.t. integrated circuits for pulse regenerator", Electron Lett., 14 (10), 294-5 (1978), EEA81-33379.
1856. Ozawa, O. and Iwasaki, H., "A vertical FET with self-aligned ion-implanted source and gate regions", IEEE Trans. Electron Devices, ED-25 (1), 56-7 (1978), EEA81-19495.
1857. Porter, J. A., "JFET-transistor yields device with negative resistance", IEEE Trans. Electron Devices, ED-23 (9), 1098-9 (1976), EEA79-46298.
1858. Vodicka, V. W. and Zuleeg, R., "Ion implanted gallium arsenide enhancement mode JFET's", Tech. Dig. - Int. Electron Devices Meet., 1975, p. 625-8, CA85-152837.
1859. Yang, P. O., "Double implanted V-MOS", International Electron Devices Meeting, Technical Digest, 1976, p. 577, EEA80-41590.

6. Resistors

1860. Antipov, I., "Forming semiconductor devices having ion implanted and diffused regions", Patent USA 4155778, Publ. May 1979.

1861. Ballatore, D., "Four-gain amplifier using ion-implanted resistors", IBM Tech. Disclosure Bull., 20 (8), 3166-7 (1978), EEA82-3643.

1862. Basso, J. E. and Nagarajan, A., "Test structure for ion-implanted resistors", IBM Tech. Disclosure Bull., 21 (1), 173-4 (1978), EEA82-17402.

1863. Beanland, D. G., Temple, W. and Chivers, D. J., "Sheet resistance variations of phosporus implanted silicon at elevated temperatures", Solid-State Electron., 21 (2), 357-60 (1978), EEA81-23876.

1864. Beck, H. G. E., "1/f noise measurements in ion-implanted silicon resistors as a function of the substrate reverse bias voltage", Solid-State Electron., 20 (12), 951-4 (1977), EEA81-14636.

1865. Beck, H. G. E., "Model of 1/f noise in ion-implanted resistors as a function of the resistance, determined by a reverse bias voltage", Solid-State Electron., 22 (5), 457-8 (1979), EEA82-38259.

1866. Bergeron, D. L. and Hansen, H. H., "Electrical characteristics and modeling of ion-implanted resistors", 1976 IEEE International Solid-State Circuits Conference, Digest of Technical Papers, p. 208-9, EEA79-45826.

1867. Bergeron, D. L. and Stephens, G. B., "Ion-implanted resistor and junction field-effect transistor process", IBM Tech. Disclosure Bull., 21 (4), 1549 (1978), EEA82-17507.

1868. Blasques, G., Hsieh, K. C., Chenette, E. R. and Van Der Ziel, A., "Flicker noise theory of diffused and ion implanted resistors", Solid-State Electron., 20 (4), 382-3 (1977), EEA80-20682.

1869. Chanclou, R., Hornung, R. and Tremintin, B., "Implanted resistors made with an additional neutral ion implantation", IBM Tech. Disclosure Bull., 22 (4), 1501-2 (1979), EEA83-14262.

1870. Chaudhari, P. K., Nelson, G. R. and Nagarajan, A., "Ion-implanted p-resistor reliability", IEEE Trans. Components, Hybrids & Manuf. Technol., CHMT-3 (2), 258-61 (1980), EEA83-44191.

1871. Das, C. J. M. and Sansen, W. M. C., "Design of ion implanted resistors with low 1/f noise", Microelectron. J., 11 (3), 24-7 (1980), EEA83-48918.

1872. Donnelly, J. P. and Hurwitz, C. E., "High-resistivity layers in n-InP produced by Fe ion implantation", Solid-State Electron., 21 (2), 475-8 (1978), EEA81-28423.

1873. Drum, C. M. and Ray, F. D., "Ion implantation of resistors for low power TTL SBC integrated circuits", Electrochemical Society Spring Meeting, Extended Abstracts, 1976, p. 346-8, EEA79-46515.

1874. Edwards, R. D., Prokop, G. S. and Rathore, H. S., "Method to improve yield of the array chips which use ion implanted resistors", IBM Tech. Disclosure Bull., 23 (6), 2298-9 (1980).

1875. Hornung, R., Nuez, J. P., Ohora, D. and Tremintin, B., "Process of forming ion-implanted resistors with implanted contact regions without an additional mask", IBM Tech. Disclosure Bull., 21 (12), 4880-1 (1979), EEA83-881.

1876. Hornung, R. and Tremintin, B., "Direct of implanted contacts in ion-implanted resistors", IBM Tech. Disclosure Bull., 21 (11), 4542-5 (1979), EEA83-738.

1877. Hsieh, K. C., McNamara, D. A., Chenette, E. R. and van der Ziel, A., "Noise in ion implanted integrated circuit resistors", Solid-State Electron., 20 (2), 164-5 (1977), EEA80-17272.

1878. Kasabov, I. D., Petrov, I. N., Nencheva, G. P. and Georgiev, V. K., "High value junction microresistors prepared by double ion implantation of ^{11}B and of noble gas ions", Bulg. J. Phys., 3 (2), 166-73 (1976), EEA83-9126.

1879. Kassabov, J. D., Voutov, M. P., Velchev, N. B. and Petrov, I. N., "Ion-implanted p-type high-resistivity layers with high temperature treatment", C. R. Acad. Bulg. Sci., 29 (4), 481-3 (1976), PA79-76621.

1880. Ku, S. M., "Boron-implanted silicon resistors", Solid-State Electron., 20 (10), 803-12 (1977), EEA81-1201.

1881. Ku, S. M. and Chu, W. K., "Some annealing properties of low-energy-antimony-implanted silicon resistors", Solid-State Electron., 22 (8), 719-22 (1979), EEA83-871.

1882. Lee, L. H., Nagarajan, A. and Smith, R. G., "Forming semiconductor devices having ion-implanted resistor and diffused regions", IBM Tech. Disclosure Bull., 21 (4), 1441-2 (1978), EEA82-17602.

1883. MacIver, B. A., "Comparison of boron and neon damage effects in boron ion-implanted resistors", Electron. Lett., 11 (20), 484-5 (1975), EEA79-4748.

1884. Nagarajan, A., "Ion implantation process for fabricating high-valued resistors", IBM Tech. Disclosure Bull., 21 (10), 4036-7 (1979), EEA82-47222.

1885. Ohzone, T., Hirao, T., Horiuchi, S. and Hozumi, H., "The features of resistor load MOS IC using the ion-implanted polysilicon resistors", Trans. Inst. Electron. and Commun. Eng. Jpn. Sect. E, E63 (4), 267-74 (1980), EEA83-44227.

1886. Petrov, I. N., "High-resistivity microresistors realized by ion implantation", Bulg. J. Phys., 5 (1), 76-85 (1978), EEA81-42068.

1887. Petrov, I. N., Nencheva, G. P., Glukharov, T. K. and Tsolov, O. P., "Layers of p-type resistivity in n-type silicon obtained by implantation of $^{49}BF_2$ ionized molecules", Bulg. J. Phys., 4 (2), 190-6 (1977), EEA81-19258.
1888. Petrov, I. N. and Petrov, Y. R., "Potential and field distributions in high-resistivity ion-implanted p-type microresistors (for MOS ICs)", Bulg. J. Phys., 6 (1), 56-70 (1979), EEA82-34375.
1889. Putney, Z. C., "Simultaneous formation of high sheet resistance ion-implanted resistors and impurity gettering for bipolar transistors", IBM Tech. Disclosure Bull., 20 (10), 4011-12 (1978), EEA82-4423.
1890. Runge, H. and Krimmel, E. F., "Influence of deep levels on the temperature coefficient of ion-implanted resistive layers in silicon", Appl. Phys., 10 (2), 181-85 (1976).
1891. Stellrecht, H. H., "Ion implanted resistor having controlled temperature coefficient", Patent USA 3947866, Publ. March 1976, CA84-172821.
1892. Toyoda, N., Niikura, I., Shimura, Y., Hozuki, T., Sugibuchi, H., Mihara, M. and Hara, T., "Ion implanted GaAs varactor diodes: capacitance uniformity", Electron. Lett., 14 (5), 152-4 (1978), EEA81-23939.
1893. Velchev, N. B. and Kovachev, V. T., "Silicon p-type microresistors at helium temperature and in a strong magnetic field", C. R. Acad. Bulg. Sci., 30 (1), 21-4 (1977), Russian, EEA80-32380.
1894. Whight, K. R., Blood, P. and Nicholas, K. H., "Implanted high value resistors", Solid-State Electron., 19 (12), 1021-7 (1976), EEA80-6588.
1895. Wiedeburg, K. H., Betz, H. and Kranz, H., "Hall effect and sheet resistivity measurements on ion-bombarded InSb", Phys. Status Solidi A, 31 (1), K69-71 (1975), EEA79-768.

7. Detectors

1896. Ajmera, P. K., "Electrical and optical studies of ion-implanted infrared sensor diodes in epitaxially grown gallium(x) indium(1-x) arsenide", Thesis, North Carolina State University at Raleigh, 1975, 128 pp., Order No. 76-0086.
1897. Betz, H., Wideburg, K., Ryssel, H. and Kranz, H., "Ion implanted InSb photodetectors", 1977 International Electron Devices Meeting, Technical Digest, p. 551-4, EEA81-30626.
1898. Chamberlain, S. G., "New profiled silicon photodetector for improved short-wavelength quantum efficiency", J. Appl. Phys., 50 (11), pt. 1, 7228-31 (1979), EEA83-14810.

1899. Deutscher, M., Drechsler, L., Matthaei, J. and Otto, G., "Effective dead layer thickness of phosphorus-implanted silicon detectors depending on the annealing temperature", Zentralinst. Kernforsch., Rossendorf Dresden, ZfK-350, 172-3 (1978), German, CA89-207939.
1900. Hubbard, G. S., Haller, E. E. and Hansen, W. L., "Ion implanted n-type contact for high-purity germanium radiation detectors", IEEE Trans. Nucl. Sci., NS-24 (1), 161-4 (1977), EEA80-26622.
1901. Mills, I. M. N., "Properties of ion-implanted and diffused photodetectors of germanium and germanium-silicon alloys", Report AD-A080412-0, Air Force Inst. of Tech., Wright-Patterson AFB, OH, 1979, 170 pp.
1902. Riepe, G. and Protic, D., "High-purity germanium detectors with both contacts made by ion-implantation", Nucl. Instrum. and Methods, 165 (1), 31-4 (1979), EEA82-49560.
1903. Ryssel, H., Lang, G., Biersack, J. P., Mueller, K. and Krueger, W., "Ion implantation doping of $Cd_{0.2}Hg_{0.8}Te$ for infrared detectors", IEEE Trans. Electron Devices, ED-27 (1), 58-62 (1980), EEA83-40310.
1904. Sueva, D., Amov, B., Dzhakov, A. and Chikov, N., "Semiconducting detectors of alpha-particles fabricated by ion implantation", Bulg. J. Phys., 4 (5), 489-95 (1977), PA81-33373.
1905. Trbojevic, D., Perovic, B., Nikolic, P. and Cvekic, V., "Photovoltaic detectors in $Pb_{0.97}Hg_{0.03}Te$ produced by Sb^+ ion implantation", Infrared Phys., 20 (3), 193-8 (1980), EEA83-41564.
1906. Vinogradova, V. G., Eremin, N. V. and Kozei, I. V., "Detectors sensitive to the point of incidence and produced by the ion implantation method", Instrum. and Exp. Tech. Inetak., 21 (5), pt. 1, 1203-6 (1978), PA83-23247.

8. CCD

1907. Bentini, G., "Ion implantation in semiconductors", Tecnol. Elettr., no. 12, 52-7 (1977), Italian, EEA81-38128.
1908. Carballo, R. A., Dockerty, R. C. and Garbarino, P. L., "Optimizing surface channel CCD performance", IBM Tech. Disclosure Bull., 22 (3), 600-1 (1979), EEA83-9910.
1909. Chen, C. H., "The charge coupled device as an interdigitally distributed RC network", Solid-State Electron., 21 (10), 1231-7 (1978).
1910. Dockerty, R. C., Pliskin, W. A., Poponiak, M. R., Revitz, M. and Shepard, J. F., "Increased hybrid breakdown strengths for CCD devices using two levels of polycrystalline silicon", IBM Tech. Disclosure Bull., 22 (5), 1849 (1979), EEA83-19117.
1911. Henderson, R. C., Mayer, D. C. and Nash, J. G., "Issues in fabricating electron devices with submicrometer dimensions", J. Vac. Sci. and Technol., 16 (2), 268-0 (1979), EEA82-38201.

1912. Herman, G. C., Hartgring, C. D. and Kleefstra, M., "Calculation of potential profiles in the junction charge-coupled device", IEEE Trans. Electron Devices, ED-25 (7), 845-7 (1978), EEA81-42419.

1913. Huang, J. S. T., "On the design of ion implanted buried channel charge coupled devices (BCCDs)", Solid-State Electron., 20 (8), 665-9 (1977), EEA80-41504.

1914. Knauer, K., "CCD versus BBD mode for CCD with ion-implanted gaps", IEEE Trans. Electron Devices, ED-24 (3), 286-8 (1977), EEA80-13516.

1915. Kohyama, S., Hatano, H., Tanaka, T. and Kubota, N., "A new multiplexed electrode-per-bit structure for a 64Kbit charge-coupled-device memory", IEEE J. Solid-State Circuits, SC-12 (4), 335-43 (1977), EEA81-10394.

1916. Lakhoua, N. and Poirier, R., "Trap levels in p and n buried channel CCD", Solid-State Electron., 21 (7), 994-7 (1978), EEA83-49064.

1917. Miyamoto, Y., Sei, H. and Ohtsuki, O., "Meander-channel charge-coupled devices", Fujitsu Sci. and Tech. J., 13 (3), 85-98 (1977), EEA81-14755.

1918. Moeller, A., "Manufacturing technology and physics of semiconductor memory elements", NTG-Fachber., 58, 79-88 (1977), German, EEA81-14627.

9. $Si-SiO_2$ Structures

1919. Aberg, A. T., "Low-temperature annealing of Al-implanted Si-SiO_2 structures", Phys. Status Solidi A, 42 (2), 639-45 (1977), EEA80-41353.

1920. Benc, I., Husa, V., Ladnar, J., Urbanec, J. and Kriz, J., "Properties of the SiO_2-Si interface after ion implantation and annealing", Proceedings of the International Conference on Ion Beam Modification of Materials, 1978, p. 137-46.

1921. Desalvo, A., Galloni, R., Rosa, R. and Zignani, F., "Experimental and computer analysis of P^+-ion penetration tails in a SiO_2-Si two-layer system", J. Appl. Phys., 51 (4), 1994-7 (1980).

1922. Didenko, P. O.. Litovchenko, V. G., Marchenko, R. I. and Romanova, G. F., "Investigating the structural changes in laminar ion-implanted Si-SiO_2 systems by MSSI method", Poluprovodn. Tekh. Mikroelektron., no. 31, 38-42 (1980), Russian, EEA84-1242.

1923. Dubchak, A. P., Lysenko, V. S., Linnik, L. G. and Nazarov, A. N., "Effect of low-energy bombardment with He^+ ions on electrophysical properties of the Si-SiO_2 interface", Ukr. Fiz. Zh., 21 (7), 1112-17 (1976), Russian, PA79-76155.

1924. Dubchak, A. P., Lysenko, V. S., Nazarov, A. N., Atamanenko, V. T., Lokshin, M. P. and Polishchuk, Y. Y., "Studies in the dielectric-semiconductor interface by the method of thermo-stimulated charge release after interposition of B^+ ions", Ukr. Fiz. Zh., 21 (7), 1197-202 (1976), Russian, PA79-76156.
1925. Gray, P. V., "Surface characterization of ion-implanted Si-SiO_2 structures", International Electron Devices Meeting, Technical Digest, 1976, p. 340-4, EEA80-36171.
1926. Hayashi, T., Okamoto, H. and Homma, Y., "Formation of abrupt interfaces between surface silicon and buried SiO_2 layers by very high dose oxygen-ion implantation", Jpn. J. Appl. Phys., 19 (5), 1005-6 (1980), EEA83-34805.
1927. Hess, D. W. and Learn, A. J., "Effects of oxidation and nitrogen annealing on ion-implantation-induced interface states in the silicon-silicon dioxide system", J. Appl. Phys., 48 (2), 834-6 (1977), EEA80-20846.
1928. Hirao, T., Inoue, K., Fuse, G., Takayanagi, S. and Yaegashi, Y., "The concentration profiles of the recoil implanted oxygen in Si after ion implantations into SiO_2-Si substrates", Radiat. Eff., 47 (1-4), 95-8 (1980), EEA83-44011.
1929. Hirao, T., Inoue, K., Yaegashi, Y. and Takayanagi, S., "The concentration profiles of phosphorus, arsenic and recoiled oxygen atoms in Si by ion implantation into SiO_2-Si", Jpn. J. Appl. Phys., 18 (3), 647-56 (1979), EEA82-26750.
1930. Izumi, T., Satoh, M., Kawamura, M. and Matsumori, T., "ESR of residual defects in silicon from recoil implantation of oxygen in Si-SiO_2 systems", Proceedings of the International Conference on Ion Beam Modification of Materials, 1978, p. 83-92.
1931. Kucirkova, A. and Stepanek, V., "Optical study of implanted SiO_2-Si system", Scr. Fac. Sci. Nat. Univ. Purkynianae Brunensis Phys., 8 (2), 69-80 (1978), EEA82-12539.
1932. Lysenko, V. S., Sytenko, T. N., Zimenko, V. I., Tyagulskii, I. P. and Snitko, O. V., "Energy spectra of shallow traps in Si-SiO_2 structures implanted with boron ions at various implantation energies", Phys. Status Solidi A, 59 (1), 115-19 (1980), EEA83-39699.
1933. Mittenbacher, J. and Gartner, K., "Proton ranges in silicon and in Si-SiO_2 double layers", Proceedings of the International Conference on Ion Beam Modification of Materials, 1978, p. 1661-72.
1934. Nedev, I. and Peikov, P., "Effect of the ^{11}B and ^{19}F ion implantation on the charge density at the Si-SiO_2 interface", Bulg. J. Phys., 5 (3), 300-4 (1978), EEA82-8261.
1935. Ragaie, H. F., "Electrical conduction in metal implanted SiO_2/Si", Electron. Lett., 16 (14), 565-6 (1980).

10. Solar Cells

1936. Bell, R. O., Ho, C. T., Ravi, K. V., Wald, R. V., Muller, J. C., Siffert, P., Van Overstraeten, R. and Palz, W., "Ion implanted solar cells from EFG silicon ribbons", 2nd Photovoltaic Solar Energy Conference, 1979, p. 153-61, EEA83-17440.

1937. Cembali, F., Galloni, R., Lotti, R., Pedulli, L. and Zignani, F., "Ion implanted, thermally and laser annealed silicon solar cells", Proceedings of the International Conference on Ion Beam Modification of Materials, 1978, p. 858-70.

1938. Christensen, O. and Bay, H. L., "Production of solar cells by recoil implantation", Appl. Phys. Lett., 28 (9), 491-4 (1976), PA79-62344.

1939. Douglas, E. C. and D'Aiello, R. V., "A study of the factors which control the efficiency of ion-implanted silicon solar cells", IEEE Trans. Electron Devices, ED-27 (4), 792-801 (1980), EEA83-42718.

1940. Eguren, J., Del Alamo, J. and Luque, A., "Optimization of p^+ doping level of n^+-p-p^+ bifacial BSF solar cells by ion implantation", Electron. Lett., 16 (16), 633-4 (1980), EEA83-52367.

1941. Fan, J. C. C., Chapman, R. L., Donnelly, J. P., Turner, G. W. and Bozler, C. O., "Ion-implanted laser-annealed GaAs solar cells", Appl. Phys. Lett., 34 (11), 780-2 (1979), EEA82-41347.

1942. Finetti, M., Ostoja, P., Solmi, S. and Soncini, G., "Minority carrier lifetime degradation during ion implanted silicon solar cell processing", Rev. Phys. Appl., 13 (12), 809-13 (1978), EEA82

1943. Gu, J., Cha, W., Gamo, K. and Namba, S., "Properties of niobium superconducting bridges prepared by electron-beam lithography and ion implantation", J. Appl. Phys., 50 (10), 6437-42 (1979), EEA82-10054.

1944. Hwang, H. L., Liu, D. C., Tang, R. S., Kao, Y. R. and Loferski, J. J., "Ion implanted grating type Si solar cells: junction depth dependence", Conf. Rec. IEEE Photovoltaic Spec. Conf., 14th, 1980, p. 381-5, EEA83-52426.

1945. Hwang, H. L., Tang, R. S., Loferski, J. J. and Yang, Y.-C., "Ion implanted grating type Si solar cells", Jpn. J. Appl. Phys., 19, suppl. 19-1, 527-32 (1980), EEA83-52379.

1946. Katzeff, J. S., "Ion-implanted laser annealed silicon solar cells", Proc. Soc. Photo-opt. Instrum. Eng., Laser Applications in Materials Processing, Vol. 198, 1979, p. 42-8, EEA83-52389.

1947. Kirkpatrick, A. R. and Melas, A. A., "Amorphous silicon solar cells by hydrogen implantation", Report SAN-3042-2, Spire Corp., Bedford, MA, 1979, 19 pp.

1948. Kirkpatrick, A. R., Minnucci, J. A. and Greenwald, A. C., "Low-cost ion implantation and annealing technology for solar cells", Conf. Rec. IEEE Photovoltaic Spec. Conf., 14th, 1980, p. 820-4, EEA83-47379.

1949. Kirkpatrick, A. R., Minnucci, J. A., Greenwald, A. C. and Josephs, R. H., "Production technology for high efficiency ion implanted solar cells", Thirteenth IEEE Photovoltaic Specialists Conference-1978, p. 706-10, EEA82-11234.
1950. Kirkpatrick, A. R., Minnucci, J. A., Shaughnessy, T. S. and Greenwald, A. C., "Silicon solar cells by ion implantation and pulsed energy processing", Twelfth IEEE Photovoltaic Specialists Conference 1976, p. 299-302, EEA81-22484.
1951. Krimmel, E. F., Runge, H., Lutsch, A. G. K., Moolman, J. S. H. and Swart, P. L., "Energy saving production of silicon single crystal solar cells by ion implantation and laser annealing", Microelectron. J., 11 (3), 28-30 (1980), EEA83-52387.
1952. Lee, C. C. and Whang, C. N., "A study on the characteristics of Si by Li^+ ion implantation and its thin film solar cells", New Phys., 16 (2), 83-8 (1976), Korean, EEA80-24922.
1953. Magee. T. J., Gibbons, J., Bean, J., Peng, J. and Armistead, R. A., "Ion implantation in cadmium telluride", Report AD-A030153/1SL, Stanford Research Inst., Menlo Park, Calif., 1976, 121 pp.
1954. Minnucci, J. A., Kirkpatrick, A. R. and Matthei, K. W., "Tailored emitter, low-resistivity, ion-implanted silicon solar cells", IEEE Trans. Electron Devices, ED-27 (4), 802-7 (1980), EEA83-42719.
1955. Muller, J. C., Grob, A., Grob, J. J., Stuck, T. and Siffert, P., "Low cost ion implantation procedure for the realization of silicon solar cells in a continuous way", Thirteenth IEEE Photovoltaic Specialists Conference-1978, p. 711-16, EEA82-11235.
1956. Muller, J. C., Ponpon, J. P., Grob, J. J., Grob, A., Stuck, R. and Siffert, P., "Low cost ion implantation for polycrystalline silicon solar cells", Proceedings of the International Conference on Photovoltaic Solar Energy, 1977, p. 897-909, EEA81-41149.
1957. Pai, Y. P. and Lin, H. C., "Analysis of ion-implanted solar cells", Thirteenth IEEE Photovoltaic Specialists Conference-1978, p. 661-4, EEA82-11226.
1958. Pai, Y. P., Lin, H. C. and Peckerar, M. C., "Current transport in ion-implanted MIS solar cells", 1977 International Electron Devices Meeting, Technical Digest, p. 51-4, EEA81-26844.
1959. Pai, Y. P., Lin, H. C., Peckerar, M. and Kocher, R. L., "Ion implanted Schottky barrier solar cell", International Electron Devices Meeting, Technical Digest, 1976, p. 470-3, EEA80-23626.
1960. Ponpon, J. P. and Siffert, P., "Silicon solar cells made by ion implantation and glow discharge", 11th IEEE Photovoltaic Specialists Conference, 1975, p. 342-8, EEA79-50698.
1961. Varma, R. and Zwerdling, S., "Merits of ion-implantation process in conjunction with appropriate annealing procedure for fabrication of silicon solar cells", Twelfth IEEE Photovoltaic Specialists Conference 1976, p. 293-8, EEA81-22483.

1962. White, C. W., "Characterization of ion-implanted, laser annealed silicon for solar cell applications", Trans. Am. Nucl. Soc., 33, 246-8 (1979), EEA83-38198.
1963. Young, R. T., White, C. W., Narayan, J., Westbrook, R. D., Wood, R. F. and Christie, W. H., "Solar cells from laser-annealed ion-implanted silicon", Thirteenth IEEE Photovoltaic Specialists Conference-1978, p. 717-23, EEA82-11236.
1964. Young, R. T. and Wood, R. F., "Solar cells from ion-implanted, laser annealed cast polycrystalline silicon", Conf. Rec. IEEE Photovoltaic Spec. Conf., 14th, 1980, P. 214-7, EEA83-47331.
1965. Zignani, F., Cembali, F., Galloni, R., Pedulli, L. and Servidori, M., "Study of ion implanted n^+ layers for silicon solar cells", Proceedings of the International Conference on Photovoltaic Solar Energy, 1977, p. 113-24, EEA81-36776.

11. Bubble Devices

1966. Ahn, K. Y., Frisch, M. and Ziegler, J. F., "Method for determining ion implantation concentrations in magnetic bubble devices", IBM Tech. Disclosure Bull., 22 (88), 3895 (1980), EEA83-29490.
1967. Ahn, K. Y. and Kane, S. M., "Toward a single-mask processing of ion-implanted bubble devices", IEEE Trans. Magn., MAG-15 (6), 1648-50 (1979), EEA83-19291.
1968. Ahn, K. Y. and Keefe, G. E., "Effects of annealing on propagation in ion-implanted contiguous-disk bubble devices", IEEE Trans. Magn., MAG-16 (4), 553-7 (1980), EEA83-44346.
1969. Ahn, K. Y., Penebre, N. A. and Ziegler, J. F., "Selective modification of magnetic properties via ion implantation", IBM Tech. Disclosure Bull., 21 (4), 1706-7 (1978), EEA82-22455.
1970. Algra, H. A. and Robertson, J. M., "The effects of ion implantation on La, Ga: YIG films as observed by spin wave resonance", J. Appl. Phys., 51 (7), 3821-6 (1980), PA83-97145.
1971. Beaulieu, T. J., Brown, R. R., Calhoun, B. A., Hsu, T. and Malozemoff, A. P., "Wall states in ion-implanted garnet films", AIP Conf. Proc., no. 34, 138-43 (1976), EEA80-9609.
1972. Blank, S. L., Wolfe, R., Luther, L. C., Lecraw, R. C., Nelson, T. J. and Biolsi, W. A., "Design and development of single-layer, ion-implantable small bubble materials for magnetic bubble devices", J. Appl. Phys., 50 (3), pt. 2, 2155-60 (1979).
1973. Dekker, P. and Deluca, J. C., "Ion-implanted bubble generator", IBM Tech. Disclosure Bull., 20 (3), 1256-7 (1977), EEA81-19649.
1974. Dove, D. B. and Schwarzl, S., "Domain-wall behavior in ion-implanted garnet layers for 1-μm bubble devices", J. Appl. Phys., 50 (9), 5906-13 (1979), EEA83-951.
1975. Deutsch, A. and Kryder, M. H., "Investigation of long-term propagation margins in ion-implanted contiguous-disk bubble devices", IEEE Trans. Magn., MAG-16 (2), 416-23 (1980).

1976. Gallagher, T. J. and Humphrey, F. B., "Velocity asymmetry during stripe head propagation in ion-implanted bubble garnet films", J. Appl. Phys., 50 (11), 7856-8 (1979), PA83-39834.
1977. Giess, E. A., Keefe, G. E. and Lin, Y. S., "Ion implanted bubble propagation structure", Patent USA 4070658, Publ. January 1978.
1978. Hu, H. L., Keefe, G. E., Ko, W. C., Kobliska, R. J., Lin, Y. S. and Russell, J. L., "Bubble devices with ion implanted underlays", IBM Tech. Disclosure Bull., 20 (11), 5032-4 (1978)
1979. Ilyashenko, E. I. and Yurchenko, S. E., "Behaviour of magnetic bubble domains in ion-implanted iron garnet films", Sov. Phys. Solid State, 18 (4), 635-6 (1976), EEA80-6804.
1980. Jouve, H., "Specific properties of ion implanted bubble circuits", J. Appl. Phys., 50 (3), pt. 2, 2246-51 (1979), EEA82-38601.
1981. Jouve, H. and Pulchaska, I. B., "Some characteristics of ion-implanted bubble chips", IEEE Trans. Magn., MAG-15 (3), 1016-20 (1979), EEA82-34471.
1982. Jouve, H., Segalini, S. and Piaguet, J., "Bubble propagation by disks formed by ion-implantation", IEEE trans. Magn., MAG-12 (6), 660-2 (1976), EEA80-9680.
1983. Ju, K. and Humphrey, F. B., "Gradientless propulsion for S=1 bubble in ion-implanted bubble garnets", J. Appl. Phys., 50 (3), pt. 2, 2212 (1979), EEA82-38524.
1984. Ju, K., Schwenker, R. O. and Hu, H. L., "Damage in garnet films produced by multiple ion implantation", IEEE Trans. Magn., MAG-15 (6), 1658 (1979), EEA83-19224.
1985. Kinsner, W. and Mikhail, S. H., "Stability of magnetic bubbles in ion implanted channels", J. Appl. Phys., 49 (3), pt. 2, 1854 (1978), EEA81-38446.
1986. Komenou, K., Hirai, I., Asama, K. and Sakai, M., "Crystalline and magnetic properties of an ion-implanted layer in bubble garnet films", J. Appl. Phys., 49 (12), 5816-22 (1978), EEA82-17709.
1987. Konishi, S., Hsu, T. and Brown, B. R., "Effect of ion-implantation on bubble translation velocity", J. Appl. Phys., 49 (3), pt. 2, 1894-6 (1978), EEA81-38392.
1988. Lin, Y. S., "Ion-implanted bubble domain devices", IBM Tech. Disclosure Bull., 20 (10), 4207-8 (1978), EEA82-4575.
1989. Lin, Y. S., "Ion implanted bubble devices", IBM Tech. Disclosure Bull., 20 (11), 5031 (1978).
1990. Maartense, I. and Searle, C. W., "Susceptibility measurements on the ion-implanted layer of bubble garnet films", Appl. Phys. Lett., 34 (1), 115-17 (1979), EEA82-22411.
1991. Mada, J. and Asama, K., "FMR study of an ion-implanted layer on bubble garnet films", J. Appl. Phys., 50 (9), 5914-19 (1979), EEA83-952.

1992. Markelis, A. V., Antonov, A. V. and Pranyavichyus, L. I., "Effect on bubble domain properties of radiation damage to iron garnet films subjected to ion implantation", Sov. Phys.-Solid State, 20 (2), 212-14 (1978), EEA81-50006.
1993. Markelis, A. V. and Pranevicius, L. I., "Hard magnetic bubble suppression mechanism in ion-implanted ferrite garnet films", Sov. Phys.-Collect., 18 (5), 62-5 (1978), EEA83-19230.
1994. Mazzeo, N. J., Sanders, I. L., Kryder, M. H. and Deutsch, A., "Stretching and sensing of 1-μm bubbles in ion-implanted contiguous disk devices", IEEE Trans. Magn., MAG-15 (6), 1905-10 (1979), EEA83-14602.
1995. Nelson, T. J., Wolfe, R., Anderson, A. W., Caruso, R., Johnson, W. A., Michaelis, P. C., Roman, B. J. and Vella-Coleiro, G. P., "Functional test circuits employing ion-implanted propagation patterns", IEEE Trans. Magn., MAG-15 (6), 1657 (1979), EEA83-19294.
1996. Nelson, T. J., Wolfe, R., Blank, S. L., Bonyhard, P. I., Johnson, W. A., Roman, B. J. and Vella-Coleiro, G. P., "Design of bubble device elements employing ion-implanted propagation patterns", Bell System Tech. J., 59 (2), 229-57 (1980), EEA83-29480.
1997. Nelson, T. J., Wolfe, R., Blank, S. L. and Johnson, W. A., "Reliable propagation of magnetic bubbles with 8 μm period ion implanted propagation patterns", J. Appl. Phys., 50 (3), pt. 2, 2261-3 (1979), EEA82-38528.
1998. North, J. C., Wolfe, R. and Nelson, T. J., "Applications of ion implantation to magnetic bubble devices", J. Vac. Sci. and Technol., 15 (5), 1675-84 (1978), EEA82-8471.
1999. Obokata, T., Uchishiba, H. and Asama, K., "Effects of in-plane magnetic field on bubble properties in ion-implanted garnet films", IEEE Trans. Magn., MAG-13 (5), 1181-3 (1977), EEA80-41734.
2000. Obokata, T., Uchishiba, H. and Asama, K., "Threefold symmetrical bubble properties in ion-implanted garnet films subjected to a rotating in-plane magnetic field", Fujitsu Sci. and Tech. J., 14 (1), 85-102 (1978), EEA81-33525.
2001. Omaggio, J. P. and Wigen, P. E., "Effects of ion implantation on the uniaxial anisotropy energy of a magnetic bubble film", J. Appl. Phys., 50 (3), pt. 2, 2264-6 (1979), EEA82-38529.
2002. Picone, P. J. and Morrish, A. H., "Magnetic structure of ion-implanted bubble garnets", Solid State Commun., 34 (9), 743-7 (1980), EEA83-40110.
2003. Pranevicius, L., "Observation of bubbles in ion implanted vacuum deposited thin films", Colloq. Eur. Surf. Vide Metall., 1978, p. 229-34.
2004. Pranevicius, L. and Markelis, A., "Influence of ion-implantation-induced stress on the properties of magnetic bubble garnets", Thin Solid Films, 60 (1), 109-11 (1979), EEA82-47415.

2005. Puchalska, I. B., Jouve, H. and Wade, R. H., "Magnetic bubbles, walls, and fine structure in ion-implanted garnets", J. Appl. Phys., 48 (5), 2069-76 (1977), EEA80-28974.
2006. Sanders, I. L. and Kryder, M. H., "Dynamics of 1 μm bubbles in ion-implanted contiguous disk structures", J. Appl. Phys., 50 (B3), 2252-57 (1979), EEA82-38527.
2007. Skrimshire, C. P., Longworth, G. and Dearnaley, G., "Ion implantation phenomena related to hard bubble suppression in magnetic bubble YIG", J. Phys. D, 12 (11), 1951-61 (1979), PA83-11238.
2008. Smit, P. H., Algra, H. A. and Robertson, J. M., "Effect of ion-implantation on magnetic properties of a garnet, as studied by Mossbauer spectroscopy", Appl. Phys., 22 (3), 299-302 (1980), EEA83-49010.
2009. Soohoo, R. F., "Ferromagnetic resonance in ion-implanted garnet bubble films", J. Appl. Phys., 49 (3), pt. 2, 1582-4 (1978), EEA81-38410.
2010. Strocka, B., Bartels, G. and Spohr, R., "Lattice strain in garnet single crystals, caused by high-energy heavy ion irradiation", Appl. Phys., 21 (2), 141-50 (1980).
2011. Uchishiba, H., Obokata, T. and Asama, K., "Effect of in-plane magnetic field on bubble collapse field in ion-implanted garnet films", J. Appl. Phys., 48 (6), 2604-7 (1977), EEA80-36562.
2012. Washburn, H. A. and Galli, G., "Ion implant profiles in garnet films", J. Appl. Phys., 50 (B3), 2267-69, (1979), EEA82-38530.
2013. Wolfe, R. and Nelson, T. J., "Crystal symmetry effects in ion-implanted propagation patterns for magnetic bubbles: 'roof-top designs", IEEE Trans. Magn., MAG-15 (5), 1323-5 (1979), EEA83-4898.

12. Others

2014. Adler, E., "Self-aligned, ion-implanted read only memory", IBM Tech. Disclosure Bull., 23 (6), 2245-6 (1980).
2015. Akomov, Y. S., Sushkov, V. P. and Kurynnyi, V. I., "Semiconducttor matrix of light-emitting elements utilizing ion implantation and diffusion heating", Patent USA 4055443, Publ. October 1977, CA87-210264.
2016. Anderson, C. L., Dunlap, H. L., Hess, L. D. and Vaidyanathan, K. V., "Laser-annealed Si and Se implants for GaAs microwave devices", AIP Conf. Proc., no. 50, 585-90 (1978), EEA82-34285.
2017. Anderson, W. T., Jr., Dietrich, H. B., Swiggard, E. W., Lee, S. H. and Bark, M. L., "Gallium arsenide transferred-electron devices by low-level ion implantation", J. Appl. Phys., 51 (6), 3175-7 (1980), EEA83-48750.
2018. Baier, H., Hinkel, H., Kraus, G. and Schmid, G. E., "Method of producing very thin layers by means of an ion implanter", IBM Tech. Disclosure Bull., 22 (3), 614 (1979), EEA83-8681.

2019. Barile, C. A., Brill, R. M., Forneris, J. L. and Regh, J., "Method of forming an integrated circuit region through the combination of ion implantation and diffusion steps", Patent USA 4060427, Publ. November 1977.
2020. Beyer, K. D. and Yeh, T., "Ion implantation gettering at elevated temperatures", IBM Tech. Disclosure Bull., 20 (8), 3122 (1978).
2021. Bogardus, E. H., Hu, S. M. and Poponiak, M. R., "Small diameter opening nozzles formed by ion implantation", IBM Tech. Disclosure Bull., 18 (2), 455 (1976).
2022. Chapman, G. E., Lau, S. S., Matteson, S. and Mayer, J. W., "Silicide formation by high-dose Si^+-ion implantation of Pd", J. Appl. Phys., 50 (10), 6321-7 (1979), PA83-19820.
2023. Christou, A. and Davey, J. E., "Refractory passivated ion-implanted GaAs ohmic Contacts", Report AD-D005735-6SL, Department of the Navy, Washington, DC, 1979, 6 pp.
2024. Chu, W. K. and Poponiak, M. R., "Method of reducing arsenic thermal diffusion by noble gas ion implantation", IBM Tech. Disclosure Bull., 21 (8), 3154-6 (1979), EEA82-38246.
2025. Destefanis, G. L., Gailliard, J. P. and Townsend, P. D., "Optical waveguides on $LiBO_3$ formed by ion implantation", Radiat. Eff., 48 (1-4), 63-8 (1980), EEA83-54052.
2026. Ehrlich, D. J., Osgood, R. M., Jr. and Deutsch, T. F., "Direct writing of regions of high doping on semiconductors by UV-laser photodeposition", Appl. Phys. Lett., 36 (11), 916-18 (1980), EEA83-39740.
2027. Fluss, M. J. and Smedeskjaer, L. C., "Ion implanted Na^{22} source for positron lifetime measurements", Report CONF-790437-8, Argonne National Lab., IL, Department of Energy, 1979, 4 pp.
2028. Gamo, K. and Namba, S., "Josephson junction fabrication by means of ion implantation and electron beam lithography", Radiat. Eff., 47 (1-4), 171-6 (1980), EEA83-44356.
2029. Gerasimenko, N. N. and Tseitlin, G. M., "Ion implantation in integrated optics", Proceedings of the International Conference on Ion Beam Modification of Materials, 1978, p. 1281-302.
2030. Gverdtsiteli, I. G., Guldamashvili, A. I., Zhirnov, V. D., Agrba, T. M. and Shaumyan, T. A., "High-effeciency ion-implanted semiconductor counter for neutrons", Instrum. & Exp. Tech., 22 (3), pt. 1, 700-2 (1979), EEA83-37160.
2031. Hara, K., "Surface treatment of quartz oscillator plate by ion implantation", Oyo Buturi, 47 (2), 145-6 (1978), Japanese, EEA81-33451.
2032. Hartemann, P., Couvard, P. and Desbois, D., "Ion-implanted surface-acoustic-wave guides on lithium niobate", Appl. Phys. Lett., 32 (5), 266-8 (1978), EEA81-24153.
2033. Hartley, N. E. W., "Applications for ion implantation as a surface treatment process in production engineering", Radiat. Eff., 44 (1-4), 19-30 (1979), PA83-34028.

2034. Heibei, J. and Voges, E., "Fabrication of strip optical waveguides in $LiBO_3$ by combined metal diffusion and ion implantation", Integrated and guided-wave optics technical digest, 1980, paper TUD3/1-4, EEA83-29678.
2035. Hinkel, H., Kaus, G., Kempf, J. and Schmitt, A., "Determination of target contaminations by ion implantation", IBM Tech. Disclosure Bull., 21 (11), 4573-4 (1979).
2036. Horng, C. T., Michel, A. E., Rupprect, H. S. and Schwenker, R. O., "Process for producing integrated circuit devices by ion implantation", Patent USA 4118250, Publ. October 1978.
2037. Howes, M. J. and Morgan, D. V., "Influence of thin implanted layers on the properties of transferred electron devices", Applications of Ion Beams to Materials, 1975, p. 91-5, Publ. 1976, EEA79-32993.
2038. Hubler, G. K., "Ion-implanted multilayer optical interference filter", Report AD-D005461-9SL, Department of the Navy, Washington, DC, 1978, 9 pp.
2039. Hunsperger, R. G. and Hirsch, N., "Ion implanted gallium arsenide semiconductor devices fabricated in semi-insulating gallium arsenide substrates", Patent USA 4033788, Publ. July 1977.
2040. Inada, T., Kato, S., Hara, T. and Toyoda, N., "Ohmic contacts on ion-implanted n-type GaAs layers", J. Appl. Phys., 50 (6), 4466-8 (1979), EEA82-47044.
2041. Inada, T., Ohkubo, T., Kato, S., Kitahara, M., Kanda, Y. and Hara, T., "GaAs Hall-effect devices fabricated by ion-implantation technique", Electron. Lett., 14 (16), 503-5 (1978), EEA81-45967.
2042. Ishiwara, H., Saitoh, S. and Furukawa, S., "Control of solid phase epitaxial growth in the Pd-Si system by carbon ion implantation", Jpn. J. Appl. Phys., 19 (5), 831-7 (1980), EEA83-38692.
2043. Kamins, T. I. and Rose, P. H., "Electron-beam annealing of ion-implantation damage in integrated-circuit devices", J. Appl. Phys., 50 (3), pt. 1, 1308-11 (1979), EEA82-38252.
2044. Koslin, M. E., Krycuk, G. A., Schatz, J. G., White, F. A. and Wood, G. M., "A feasibility study of ion implantation techniques for mass spectrometer calibration", Report N78-21431-9SL, NASA, Langley Research Center, Langley Station, VA, 1978, 21 pp.
2045. Land, C. E. and Peercy, P. S., "Ion-implanted PLZT for photoferroelectric image storage and display devices", Ferroelectrics, 27 (1-4), 131-6 (1980), PA83-80315.
2046. Lee, H. S., "Step-oxide ion-implanted memory", IBM Tech. Disclosure Bull., 20 (7), 2586-7 (1977), EEA81-46073.
2047. Liu, S. G., Douglas, E. C. and Wu, C. P., "High-energy ion implantation for multigigabit-rate GaAs integrated circuit", Report AD-A076237-7, RCA Labs, Princeton, NJ, 1979, 79 pp.

2048. Matteson, S., Roth, J. and Nicolet, M. -A., "Ion induced silicide formation in niobium thin films", Proceedings of the International Conference on Ion Beam Modification of Materials, 1978, p. 1255-68.
2049. Mizutani, T., Honda, T., Yamazaki, H. and Fugjimoto, M., "GaAs planar Gunn devices with sulfur-ion implanted n layers", Solid-State Electron., 20 (5), 443-5 (1977), EEA80-24912.
2050. Mizutani, T., Ishida, S., Kurumada, K. and Ohmori, M., "Selectively ion implanted Gunn-F.E.T. integrated circuits for pulse regenerator", Electron. Lett., 14 (10), 294-5 (1978), EEA81-33379.
2051. Nakatsuka, M., Tanaka, K. and Kikkawa, T., "Effects of thin conductive film mask on ion implantation", J. Electrochem. Soc., 125 (11), 1829-32 (1978).
2052. Niikura, I., Toyoda, N., Shimura, Y., Yokoyama, S., Mihara, M., Hayashi, T. and Hara, T., "GaAs varactor diodes for UHF TV tuners fabricated by ion implantation", Electron. Lett., 14 (1), 9-10 (1978), EEA81-16304.
2053. Nishimura, T., Aritome, H., Masuda, K. and Namba, S., "Optical waveguiding and electrooptic modulation in ion-implanted CdTe", Jap. J. Appl. Phys., 15 (11), 2283-4 (1976), EEA80-6896.
2054. Noonan, J. R., Kirkpatrick, C. G., Myers, D. R. and Streetman, B. G., "Design for a low temperature ion implantation and luminescence cryostat", J. Phys. E, 9 (4), 264-6 (1976), CA85-80109.
2055. Ohdomari, I., Tu, K. N. and Hammer, W., "Ion beam modification of silicide-silicon interfaces", Proceedings of the International Conference on Ion Beam Modification of Materials, 1978, p. 995-1008.
2056. Pankove, J. I. and Wu, C. P., "Photoluminescence from hydrogenated ion-implanted crystalline silicon", Appl. Phys. Lett., 35 (12), 937-9 (1979), EEA83-19449.
2057. Pankove, J. and Wu, C., "Aluminum patterning by ion implantation", 1980 International Electron Devices Meeting, Technical Digest, p. 156-8.
2058. Pham, M. T. and Hueller, J., "Ion implanted silicon-electrolyte interface", J. Appl. Electrochem., 7 (6), 531-7 (1977), CA88-14995.
2059. Pianetta, P. A., Stolte, C. A. and Hansen, J. L., "Nonalloyed ohmic contacts to electron-beam-annealed Se-ion-implanted GaAs", Appl. Phys. Lett., 36 (7), 597-9 (1980), EEA83-29056.
2060. Prager, R., "Ion implanted waveguides (integrated optics)", Wiss. Z. Friedrich-Schiller-Univ. Jena, Math. Naturwiss. Reihe, 27 (2-3), 133-4 (1978), German, EEA81-46297.
2061. Rosinski, W., "Metal-semiconductor contacts by ion implantation", Int. Conf. Ion Implantation Semicond., Proc., 1978, p. 570-3, CA90-96420.

2062. Tanoue, H., Tsurushima, T. and Kataoka, S., "GaAs Hall element fabricated by ion implantation", IEEE Trans. Electron Devices, ED-27 (6), 1188-92 (1980), EEA83-44049.

2063. Townsend, P. D., "Ion implantation and integrated optics", J. Phys. E, 10 (3), 197-203 (1977), EEA80-13901.

2064. Tu, K. N. and Van Vechten, J. A., "Ink jet silicon nozzles coated with ion implanted silicide for corrosion protection", IBM Tech., Disclosure Bull., 21 (8), 3371 (1979).

2065. Valette, S., Labrunie, G., Deutsch, J. C. and Lizet, J., "Planar optical waveguides achieved by ion implantation in zinc telluride: general characteristics", Appl. Opt., 16 (5), 1289-96 (1977), EEA80-29057.

2066. Wang, K. L., Bacon, F. and Reihl, R. F., "Induced interface interactions in Ti/Si systems by ion implantation", J. Vac. Sci. & Technol., 16 (2), 130-3 (1979), EEA82-38127.

2067. Wang. K. L., Bacon, F. and Reihl, R. F., "Interface modification of refractory metal-silicon structures by ion implantation", J. Vac. Sci. & Technol., 16 (6), 1909-12 (1979), EEA83-34695.

2068. Welch, B. M., "Method for the fabrication of gallium arsenide semiconductor devices", Report AD-D002755/7SL, Department of the Air Force, Washington, D. C., 1976, 17 pp.

2069. Welch, B. M. and Eden, R. C., "Planar GaAs integrated circuits fabricated by ion implantation", 1977 International Electron Devices Meeting, Technical Digest, p. 205-8, EEA81-28584.

2070. Welch, B. M. and Pashley, R. D., "Ion implantation method for the fabrication of gallium arsenide semiconductor devices utilizing an aluminum nitride protective capping layer", Patent USA 4058413, Publ. November 1977.

2071. Woo, D. S., "All-ion-implantation process for production of integrated circuits", Insul. Circuits, 26 (1), 39-40 (1980).

2072. Yasaitis, J. A., "Ion implantation of neon in silicon for planar amorphous isolation", Electron. Lett., 14 (15), 460-2 (1978), EEA81-42437.

2073. Zucca, R., Eisen, F. H., Welch, B. M., Eden, R. and Fairman, R. D., "Ion implanted GaAs I.C. process technology", Report AD-A052894-3SL, Rockwell International, Thousand Oaks, Calif., 1978, 67 pp.

VII. EQUIPMENT

1. Accelerator/Implantor Systems

2074. Aitken, D., "The design philosophy for a 200 kV industrial high current ion implanter", Nucl. Instrum. & Methods, 139, 125-34 (1976), EEA80-13459.

2075. Aitken, D., "The performance of the series III 200 kV high current industrial ion implanter", Radiat. Eff., 44 (1-4), 159-66 (1979), EEA83-14303.

2076. Allison, R., Jr., Church, D., Glavish, H., Noeth, A. and Schechter, D., "Design and performance of a high current 200 keV electrostatically scanned ion implantation accelerator", Electrochemical Society Spring Meeting (Extended Abstracts), 1976, p. 529, EEA79-46108.
2077. Anand, K. V., El-Dhaher, A. and Badawi, M. H., "A low-cost ion implantation system", Electron. Eng., 49 (600), 43-7 (1977), EEA81-10264.
2078. Bannenberg, J. G., "Accelerator systems for ion implantation research", Radiat. Eff., 44 (1-4), 3-10 (1979), EEA83-16118.
2079. Bannenberg, J. G. and Boerboom, A. J. H., "Design of a 320 kV ion accelerator for multiply charged heavy ions", IEEE Trans. Nucl. Sci., NS-22 (3), 1675-8 (1975), EEA78-43695.
2080. Bannenberg, J. G., Doorn, S. and Saris, F. W., "A new accelerator for multiply charged ions at low energy", International Conference on Low-Energy Ion Beams, 1977, p. 63-9, EEA81-40211.
2081. Bayer, E. H., Paul, L. F. and Kranik, J. R., "Architecture and control of a high current ion implanter system", Radiat. Eff., 44 (1-4), 167-72 (1979), EEA83-19002.
2082. Bird, H. M. B., Jackson, J. H., Weissman, B. and Williams, N., "PR-200 ion implantation system", J. Vac. Sci. & Technol., 15 (3), 1080-5 (1978), EEA81-42351.
2083. Boroffka, H., Krimmel, E. and Runge, H., "Ion implantation device", Patent German 2222736, Publ. November 1973, CA80-42207.
2084. Camplan, J., Chaumont, J., Meunier, R., Graber, R., Rouge, J. C., Stocker, R. and Wegmann, L., "A high current ion implanter with hybrid scanning", Nucl. Instrum. & Methods, 171 (2), 245-50 (1980), EEA83-39758.
2085. Chang, H. H., Han, C. -E., Wang, W. H. and Chen, J. Y., "400 keV heavy ion implanter", Radiat. Eff., 44 (1-4), 181-4 (1979), EEA83-14305.
2086. Croset, M., Dieumegard, D., Grouillet, A., Icole, I. and Morizot, M., "The 400 kV heavy ion accelerator of the Central Research Laboratory", Rev. Tech. Thomson-CSF, 9 (4), 849-60 (1977), French, EEA81-25954.
2087. Curtis, C. D., "Ion source and injector development", 1976 Proton Linear Accelerator Conference, p. 179-88, EEA80-18370.
2088. Dimmler, D. G., "Computer networks in future accelerator control systems", IEEE Trans. Nucl. Sci., NS-25 (2), 974-88 (1978), EEA81-30806.
2089. Fischer, G., Hill, A. E. and Colligon, J. S., "A two-gun-beam system for dynamic recoil implantation", Vacuum, 28 (6-7), 277-81 (1978), EEA82-8270.
2090. Freitag, K., "A facility for ion implantation in samples colder than 0.5K (for nuclear orientation studies)", Radiat. Eff., 44 (1-4), 185-90 (1979), PA83-34033.

2091. Furuno, K., Tagishi, Y., Yamanouchi, M. and Sanada, J., "A 700 keV tandem accelerator", Nucl. Instrum. & Methods, 161 (1), 9-12 (1979), EEA82-28341.

2092. Gale, D. S., "Design of a 300 kV heavy ion accelerator", Proceedings of the 4th Conference on the Scientific and Industrial Applications of Small Accelerators, 1976, p. 471-5, EEA80-38397.

2093. Hechtl, E. and Kemmer, J., "An isotope separator implantation terminal for the preparation of particle detectors", Radiat. Eff., 44 (1-4), 191-4 (1979), EEA83-19003.

2094. Ilyushkin, V. A., Kotlyarevskii, M. B., Ludzish, O. S., Noskov, D. A., Shibaev, Y. A. and Shvetsov, Y. V., "Ion-beam apparatus for semiconductor doping", Instrum. & Exp. Tech., 21 (5), pt. 2, 1447 (1978), EEA83-14291.

2095. Jackson, J. H., Bird, H. M., Flemming, J. P., Hofer, G. J., McCallum, J. G., Mostek, P. J., Robertson, G. I., Rodde, A. F., Weissman, B. and Williams, N., "High throughput ion implantation systems in Western Electric", Radiat. Eff., 44 (1-4), 59-70 (1979), EEA83-14298.

2096. Jackson, J. H., Bird, H. M., Hofer, G. J., Mostek, P. J., Robertson, G. I., Rodde, A. F. and Weissman, B., "PR-500, a high voltage, general purpose ion implantation system", Electrochemical Society Spring Meeting, Extended Abstracts, 1976, p. 523-5, EEA79-46106.

2097. Jesser, W. A., Horton, J. A. and Scribner, L. L., "Adaption of an ion accelerator to a high voltage electron microscope", Radiat. Eff., 29 (2), 79-82 (1976), CA85-115165.

2098. Keller, J. H., McKenna, C. M., Winnard, J. R., Hicks, W. W., Hoffman, J. E., Kranik, J. R. and Mueller, W. F., "Development of a prototype high-current low-energy ion implanter", Radiat. Eff., 44 (1-4), 195-200 (1979), EEA83-14306.

2099. Lawther, C., Anand, K. V. and Fakhouri, B., "A low energy implantation system for the fabrication of Mn implanted ZnS electroluminescent films", Nucl. Instrum. & Methods, 148 (2), 389-94 (1978), EEA81-19809.

2100. Martel, C. and Duchatelle, J. P., "Apparatus and method for implanting ions in substrates", Patent French 2263028, Publ. October 1975, CA84-115053.

2101. McCallum, J. G., Robertson, G. I., Rodde, A. F., Weissman, B. and Williams, N., "PR-30 ion implantation system", J. Vac. Sci. & Technol., 15 (3), 1067-9 (1978), EEA81-45935.

2102. Pasztor, E., "The KFKI 150 kV ion-implanter", Report KFKI-76-63, Hungarian Acad. Sci., Budapest, 1976, 33 pp.

2103. Paterson, J. A., Duffy, T. J., Haughian, J. M., Biagi, L. A. and Yee, D. P., "The design and fabrication of an ion accelerator for TFTR-type neutral beam systems", Proceedings of the Seventh Symposium on Engineering Problems of Fusion Research, Pt. II, 1977, p. 1421-4, EEA81-40189.

2104. Perng, C. C. and Lin, S. Y., "150 keV heavy ion generator for sputtering studies", Chin. J. Phys., 16 (1), 20-3 (1978), EEA82-2262.
2105. Primenko, G. I., Strizhak, V. I., Usikov, Y. I. and Estulin, I. V., "High-precision ion accelerator", Instrum. & Exp. Tech., 21 (5), pt. 1, 1178-80 (1978), EEA83-16115.
2106. Ribarov, S. P., "A simple linear accelerator for doping semiconductors by ion implantation technology", C. R. Acad. Bulg. Sci., 28 (4), 461-4 (1975), EEA78-39287.
2107. Robinson, R. J., "Ion implanters overcoming current barriers", Semicond. Internat., 2 (5), 45-53 (1979).
2108. Ryding, G. and Wittkower, A. B., "Industrial ion implantation machines", IEEE Trans. Manuf. Technol., MFT-4 (1), 21-31 (1975), CA84-10080.
2109. Ryding, G., Wittkower, A. B. and Rose, P. H., "Varian Extrion high dose implanters", Electrochemical Society Spring Meeting, Extended Abstracts, 1976, p. 530-1, PA79-91568.
2110. Ryding, G., Wittkower, A. B. and Rose, P. H., "Features of a high-current implanter and a medium-current implanter", J. Vac. Sci. & Technol., 13 (5), 1030-6 (1976), EEA80-6584.
2111. Talada, D. D., "Evaluation of a low energy ion implantation system", Report GE-EE-74-36, Air Force Inst. Tech., Wright-Patterson AFB, Ohio, 1974, 77 pp., EEA78-24234.
2112. Wegner, H. E., "Advances in electrostatic accelerators", IEEE Trans. Nucl. Sci., NS-22 (3), 946-55 (1975), EEA78-43629.
2113. Wittkower, A. B., "Calibration of ion implantation systems", Solid State Technol., 21 (11), 61-7 (1978), EEA82-8267.

2. Ion Source/Extractor

A. General

2114. Aberth, W., Schnitzer, R., Engesser, F. and Anbar, M., "Design of a general purpose duoplasmatron", Report LBL-3399 (Suppl.), Stanford Research Inst., Menlo Park, Calif., 1974, p. 1.7.1-I.7.4, NSA33-25611.
2115. Alton, G. D., "Ion sources for accelerators", Proc. Conf. Appl. Small Accel., 3rd, 1974, p. 11-30, Publ. 1975, CA85-53241.
2116. Alton, G. D., "A universal negative or positive ion source", IEEE Trans. Nucl. Sci., NS-26 (3), pt. 2, 3708-12 (1979), EEA82-35984.
2117. Al'tudov, Y. K., Basova, T. A., Bykovskii, Y. A., Degtyarev, V. G., Kolosov, Y. N., Laptev, I. D. and Nevolin, V. N., "Laser-plasma source for ion implantation in solids", Sov. Phys.-Tech. Phys., 24 (9), 1077-9 (1979), EEA83-41815.
2118. Amano, J. and Lawson, R. P. W., "Thin film deposition using low-energy ion beams. IV. Ion source modification", J. Vac. Sci. & Technol., 15 (1), 118-19 (1978), EEA81-30822.

2119. Anon., "Ion source", Electron. Prod., 8 (5), 33 (1979), EEA83 13472.
2120. Antonsen, T. M., Jr. and Ott, E., "Theory of intense ion beam acceleration", Phys. Fluids, 19 (1), 52-9 (1976), EEA79-22385
2121. Apard, P., Bliman, S. and Geller, R., "A multipurpose highly charged ion source", Nucl. Instrum. & Methods, 129 (2), 357-63 (1975), EEA79-10725.
2122. Aubert, J., Gautherin, G., Gilles, J. P., Grandchamp, J. P. and Lejeune, C., "Emittance and brightness of multiply-charged ion beams from duopigatron and duoplasmatron sources" IEEE Trans. Nucl. Sci., NS-23 (2), 1088-92 (1976), EEA79-26280.
2123. Aubert, J., Gautherin, G. and Lejeune, C., "The triplasmatron new positive and negative ion source", Proc. Int. EMIS Conf. Low Energy Ion Accel. Mass Sep., 8th, 1973, p. 155-62, CA85-12705.
2124. Bacon, F. M., "Gas discharge ion source. I. Duoplasmatron", Rev. Sci. Instrum., 49 (4), 427-34 (1978), EEA81-35701.
2125. Bacon, F. M., Bickes, R. W., Jr. and O'Hagan, J. B. O., "Gas discharge ion source. II. Duopigatron", Rev. Sci. Instrum., 49 (4), 435-9 (1978), EEA81-35702.
2126. Batalin, V. A., Kolomietz, A. A., Kondratjev, B. K. and Kujbida, R. P., "Duoplasmatron type source with cold cathode" IEEE Trans. Nucl. Sci., NS-23 (2), 1097 (1976), EEA79-26281.
2127. Baumann, H., Bethge, K. and Klein, G., "The emittance and brightness of ion beams from a PIG ion source with end extraction", IEEE Trans. Nucl. Sci., NS-23 (2), 1081-3 (1976), CA84-186233.
2128. Bayer, E. H., Frederick, J. J., Jung, G. and Popp, G., "Loading system for charging ion implant source chamber", IBM Tech Disclosure Bull., 21 (9), 5635-7 (1979).
2129. Beaver, E. M., "An improved electron impact ion source power supply", Mass Spectrosc., 22 (3), 203-7 (1974), EEA78-30616.
2130. Beck, S., Brack, K. and Gansauge, P., "Implantation with boron ions utilizing a boron oxide ion source", Patent USA 3960605, Publ. June 1976, CA85-86290.
2131. Bernhardi, K., Fuchs, G., Goldman, M. A., Herbert, H. C., Walcher, W. and Wiesemann, K., "An electron-cyclotron-resonance plasma as a source of multiply charged ions", IEEE Trans. Nucl. Sci., NS-23 (2), 999-1005 (1976), CA84-186218.
2132. Betigeri, M. G., Bhatia, M. S. and David, T. P., "Production of 50 keV intense heavy ion beam", Nucl. Instrum. & Methods, 128, 29-33 (1975).
2133. Bex, L., Clark, D. J., Ellsworth, C. E., Estrella, R. M., Gough, R. A. and Holley, W. R., "Emittance measurements of high charge-state argon beams from a PIG source", IEEE Trans. Nucl. Sci., NS-23 (2), 1077-80 (1976), CA84-186232.
2134. Bird, H. M. B. and Flemming, J. P., "Development of the ERC cold-cathode ion source for use on the PR-30 ion-implantation system", J. Vac. Sci. & Technol., 15 (3), 1070-5 (1978).

2135. Bogdanov, D. D., Voborzhil, I., Demyanov, A. V., Karnaukhov, V. A. and Petrov, L. A., "Ion source with surface ionization for BEMS-2 mass-separator in a heavy ion beam", Report JINR-P13-8961, Jt. Inst. Nucl. Res., Dubna, USSR, 1975, 12 pp., CA84-129462.
2136. Boskovic, D. and Subotic, K. M., "RF ion source with a 100 Mc/s oscillator", Fizika, 9, suppl. 1, 75 (1977), EEA81-12101.
2137. Briand, P., Chan-Tung, N., Geller, R. and Jacquot, B., "Triple-mafios multicharged heavy ion source", Rev. Phys. Appl., 12 (8), 1135-8 (1977), French, EEA81-7845.
2138. Briska, M., Malin, K. and Schmitt, A., "Source material for the ion implantation of arsenic and boron", IBM Tech. Disclosure Bull., 20 (3), 1048 (1977), EEA81-19424.
2139. Brown, H. L. and Cross, R. H., "Technique for measuring the natural angular divergence of ions exciting low-intensity ionization sources", Rev. Sci. Instrum., 47 (7), 887-8 (1976), PA79-85875.
2140. Burkard, K. H., Dumanski, W., Kirchner, R., Lauppe, W., Otto, H., Roeckl, E. and Seckert, F., "Ion source development", Report NP-20343, Gesellschaft fuer Schwerionenforschung mbH, Darmstadt, Germany, 1974, p. 118-24, German, NSA32-15058.
2141. Chavet, I., Kanter, M. and Menat, M., "The divergence profile of a wedge-type ion beam", Nucl. Instru. & Methods, 153 (2-3), 317-32 (1978), EEA81-47516.
2142. Ciuti, P., "Magnetic topography in duoplasmatron ion sources", Energia Nucleare, 24, 89-96 (1977), EEA80-30335.
2143. Clark, D. J., "Recent progress in heavy ion sources", IEEE Trans. Nucl. Sci., NS-24 (3), 1064-9 (1977), EEA80-43901.
2144. Clausnitzer, G., Klinger, H., Mueller, A. and Salzborn, E., "Investigation of an electron beam ion source for the production of multiply charged heavy ions", IEEE Trans. Nucl. Sci., NS-23 (2), 1027-30 (1976), CA84-186224.
2145. Coburn, J. W., Phipps, P. B. and Winters, H. F., "Ion implantation using a laser-solid interaction as the ion source", IBM Tech. Disclosure Bull., 22 (6), 2451 (1979), EEA83-23910.
2146. De Fries, R. M., "Elemental arsenic/silicon compounds for ion implant source materials, ", IBM Tech. Disclosure Bull., 18 (2), 421 (1975), EEA78-37403.
2147. De Fries, R. M., "Elemental boron/silicon compounds for ion-implant source materials", IBM Tech. Disclosure Bull., 19 (11), 4150 (1977), EEA81-10272.
2148. Dietze, H. -J., Zahn, H. and Schmidt, W., "The form of ion pulses from a laser ion-source", Int. J. Mass Spectrum. & Ion Phys., 21 (3-4), 231-40 (1976), German, EEA80-1631.
2149. Dove, D. B., Molzen, W. and Johnson, G., "Modification to enhance the beam current of a simple ion gun", Rev. Sci. Instrum., 47 (3), 299-300 (1976), EEA79-26304.
2150. Dreyfus, R. W., "CW monoenergetic Cs^+ ion source by photoionization", Appl. Phys. Lett., 36 (7), 495-7 (1980).

2151. Dreyfus, R. W. and Hodgson, R. T., "High brightness ion source", Patent USA 3914655, Publ. October 1975, NSA33-15551.
2152. D'yachkov, B. A., Zaryankin, N. M., Logunov, V. I. and Nasonov A. V., "High-frequency source of low-energy ions", Instrum. & Exp. Tech., 21 (2), pt. 2, 469-71 (1978), EEA82-24277.
2153. El-Dhaher, A. H. G., Anand, K. V. and Badawi, M. H., "RF ion source for a small ion implanter", Electron, 134, 46, 48 (1978), EEA81-42345.
2154. Fink, J. H. and Schumacher, B. W., "Characterization of charged particle beam sources", Nucl. Instrum. Methods, 130 (2), 353-8 (1975), CA84-113274.
2155. Fitch, R. K. and Venkatesh, S., "Comparison of the McClure and McIlraith type ion sources", Vacuum, 29 (1), 19-21 (1979).
2156. Flemming, J. P., "Penning source for ion implantation", J. Vac. Sci. & Technol., 12 (6), 1369-72 (1976), EEA79-20320.
2157. Forrester, A. T., Goebel, D. M. and Crow, J. T., "IBIS: a hollow cathode multiple boundary ion source", Appl. Phys. Lett. 33 (1), 11-13 (1978).
2158. Franks, J., "Properties and applications of saddle-field ion sources", J. Vac. Sci. Technol., 16 (2), 181-3 (1979), CA91-31281.
2159. Franks, J. and Ghander, A. M., "A saddle field ion source of spherical configuration for etching and thinning applications", Vacuum, 24 (10), 489-91 (1974), EEA78-34981.
2160. Freeman, J. H., Chivers, D. J. and Gard, G. A., "Ion beam studies. IV. The use of multiply-charged and polyatomic ions in an implantation accelerator", Nucl. Instrum. & Methods, 143 (1), 99-115 (1977), EEA80-30339.
2161. Freeman, J. H., Chivers, D. J., Gard, G. A. and Temple, W., "Ion beam studies. VI. The production of heavy ion beams", Nucl. Instrum. & Methods, 145 (3), 473-505 (1977), EEA81-3343.
2162. Geller, R., "Electron cyclotron resonance multiply charged ion sources", IEEE Trans. Nucl. Sci., NS-23 (2), 904-12 (1976), CA84-186216.
2163. Gerlach, R. L. and Rossini, G., "Ion beam apparatus with separately replaceable elements", Patent USA 3930163, Publ. December 1975, NSA33-24291.
2164. Green, T., "Beam optics for low extraction with a high-voltage-ratio acceleration-deceleration system", J. Phys. D, 9, 1165-71 (1976).
2165. Green, T. S., Goble, C., Inman, M. and Martin, A. R., "The measurement of ionization rates in ion sources", 7th European Conference on Controlled Fusion and Plasma Physics, Pt. I, 1975, p. 93, EEA79-18210.
2166. Grisham, L. R., Tsai, C. C., Whealton, J. H. and Stirling, W. L., "Effect of emission aperture shape upon ion optics", Rev. Sci. Instrum., 48 (8), 1037-41 (1977).
2167. Gruzdev, V. A., Kreindel, Y. E. and Semenov, A. P., "A source of metal ions", Instrum. & Exp. Tech., 21 (2), pt. 2, 471-3 (1978), EEA82-24278.

2168. Hamm, R. W., Choate, L. M. and Kenefick, R. A., "Current results from the Texas A & M electron beam ion source", IEEE Trans. Nucl. Sci., NS-23 (2), 1013-16 (1976), CA84-186221.
2169. Harper, J. M. E., Speidell, J. L., Tuttle, A. H., "Silicon ion source grids", IBM Tech. Disclosure Bull., 23 (6), 2582-3 (1980).
2170. Heinicke, E., Nickel, W., Rang, B. and Bethge, K., "Studies of a PIG source for positive ion extraction", IEEE Trans. Nucl. Sci., NS-23 (2), 1061-4 (1976), CA84-186230.
2171. Hicks, W. W. and Keller, J. H., "An accel structure for an ion implanter extractor", IBM Tech. Disclosure Bull., 20 (4), 1367-9 (1977), EEA81-23861.
2172. Hirshfield, J. L., "Laser-initiated vacuum arc for heavy ion sources", IEEE Trans. Nucl. Sci., NS-23 (2), 1006-7 (1976), EEA79-30260.
2173. Hoff, P., Jacobsson, L., Johansson, B., Aagaard, P., Rudstam, G. and Zwicky, H.-U., "Production of molecular ions in the integrated target/ion-source of an isotope-separation-on-line system", Nucl. Instrum. & Methods, 172 (3), 413-18 (1980), EEA83-51209.
2174. Hudson, E. D. and Mallory, M. L., "Heavy ion source support gas mixing experiments", IEEE Trans. Nucl. Sci., NS-24 (3), 1590-2 (1977), EEA80-43907.
2175. Hudson, E. D., Mallory, M. L. and Lord, R. S., "Production of positive ion beams from solids", IEEE Trans. Nucl. Sci., NS-23 (2), 1065-8 (1976), EEA79-26277.
2176. Hudson, E. D., Palmer, G. A., Haley, C. L. and Mallory, M. L., "Cyclotron heavy ion beam intensity enhancement by using an easily ionized support gas in the ion source", Nucl. Instrum. and Methods, 141 (2), 381-2 (1977), PA80-34490.
2177. Ikuta, K., Mohri, A. and Masuzaki, M., "Production of strong current ion beams in a non-uniform magnetic field", Report IPPJ-224, Nagoya Univ., Japan, 1975, 6 pp., EEA79-22386.
2178. Ilyushkin, V. A., Noskov, D. A., Shangin, A. S. and Shibaev, Y. A., "Multichannel ion source", Instrum. & Exp. Tech., 20 (1), pt. 2, 228-30 (1977), EEA81-16839.
2179. Ishikawa, J., Sano, F., Tsuji, H. and Takagi, T., "Ion extraction system of the Kyoto beam-plasma type ion source", International Conference on Low-Energy Ion beams, 1977, p. 84-9, EEA81-40214.
2180. Ivanov, N. S., Kabachenko, A. P., Kuznetzov, I. V. and Tarantin, N. I., "A gas-discharge ion source for an on-line electromagnetic mass separator at a heavy ion accelerator. I. Design and efficiency", Nucl. Instrum and Methods, 147 (2), 333-5 (1977), PA81-17456.
2181. Kabachenko, A. P., Kuznetsov, I. V., Lee, G. S. and Tarantin, N. I., "A gas discharge ion source for an on-line electromagnetic mass separator at a heavy ion accelerator. II. The rate of performance", Nucl. Instrum. and Methods, 147 (2), 337-40 (1977), PA81-17457.

2182. Keersmaecker, R. D., "Oxidized BN: an excellent source material for producing boron ion beams", Rev. Sci. Instrum., 46 (7), 921-2 (1975), EEA79-6383.
2183. Keller, R. and Muller, M., "Duoplasmatron development", IEEE Trans. Nucl. Sci., NS-23 (2), 1049-52 (1976), EEA79-30268.
2184. Keller, R. and Muller, M., "Duoplasmatron sputter source for multiply charged ions of solids", International Conference on Low-Energy Ion beams, 1977, p. 40-3, EEA81-40209.
2185. Keller, R. and Winter, H., "Versatile source for intense multiply-charged ion beams", Part. Accel., 7 (2), 77-82 (1976), EEA79-49104.
2186. Khirnyi, Y. M. and Kochemasova, L. N., "A source of doubly charged helium ions for an electrostatic accelerator", Instrum. & Exp. Tech., 17 (5), 1269-71 (1974), EEA78-26215.
2187. Khorossany, M. and Fitch, R. K., "Energy distribution of the ions produced by saddle field ion sources", Vacuum, 27 (3), 159-62 (1977), EEA81-3345.
2188. Kim, J. and Davis, R. C., "Enhancement of H^+ fraction in hydrogen ion sources by means of predissociation method", Appl. Phys. Lett., 30 (3), 130-2 (1977).
2189. Kirchner, R., "Plasma and beam characteristics of FEBIAD - a low-intensity, high-efficiency ion source", International Conference on Low-Energy Ion Beams, 1977, p. 29-33, EEA81-40208.
2190. Kirchner, R. and Roeckl, E., "Cathode with long lifetime for operation of ion sources with chemically agressive vapors", Nucl. Instrum. & Methods, 127, 307-9 (1975).
2191. Kirchner, R. and Roeckl, E., "Investigation of small-volume gaseous discharge ion sources for isotope separation on-line", Nucl. Instrum. Methods, 131 (2), 371-4 (1976), CA84-142281.
2192. Kloppel, K. -D., Peters, H. and Seidel, W., "Investigation of an arc discharge ion source with energy selection (CMA)", Nucl. Instrum. & Methods, 144 (3), 407-11 (1977), EEA80-43916.
2193. Knystautas, E. J. and Lapointe, R., "Triply charged argon and neon ion beams from RF sources", Nucl. Instrum. & Methods, 157 (3), 607-9 (1978), EEA82-14818.
2194. Ko, W. C. and Schien, A., "Divergence in a high-intensity ion beam from a concave multiple-aperture source electrode", Nucl. Instrum. Methods, 130 (2), 359-63 (1975), CA84-113176.
2195. Kobayashi, M. and Takagi, A., "Construction of duoplasmatron ion source for the KEK-PS", IEEE Trans. Nucl. Sci., NS-24 (3), 1587-9 (1977), EEA80-43906.
2196. Koestner, J. H. and Scheer, G. F., "Ion bombardment apparatus with removable extractor assembly", IBM Tech. Disclosure Bull. 20 (10), 3906-7 (1978), EEA82-4300.
2197. Komkov, V. S. and Kovura, Y. A., "Distribution of currents in the discharge chamber of an ion source with volume ionization", Radio Eng. & Electron. Phys., 23 (8), 127-30 (1978), EEA83-11806.
2198. Kornelsen, E. U., "Ion gun with relatively high yield at low pressure", J. Vac. Sci. & Tecnol., 13, 716-20 (1976).

2199. Lejeune, C., "Oscillations in ion-source discharges and their relation to ion beam properties", Proc. Symp. Ion Sources Form. Ion Beams, 2nd, 1974, 11, 12 pp., Publ. 1975, CA85-200887.
2200. Lejeune, C., Grandchamp, J. P., Gilles, J. P. and Aubert, J., "A c.w. duopigatron multiply-charged ion source", IEEE Trans. Nucl. Sci., NS-23 (2), 1084-7 (1976), CA84-186234.
2201. Lukner, C., Latzel, G., Paetz, H. and Schieck, G., "Matching of heavy ion sources to tandem accelerators", Nucl. Instrum. & Methods, 146 (1), 139-42 (1977), EEA81-12104.
2202. Ma, M. H., Yao, W., Pai, K. P. and Chang, T. -H., "An improved radio frequency heavy ion source (for ion implantation)", Radiat. Eff., 44 (1-4), 207-12 (1979), PA83-32561.
2203. Machet, J., "Origin of ionic oscillations in an ion source with magnetic confinement and axial extraction", Rev. Phys. Appl., 10 (6), 453-61 (1975), French, CA84-143423.
2204. Maczka, D., Gromova, I. I. and Zuk, W., "An investigation of the penning phenomenon in the plasma of a magnetron ion source", Nucl. Instrum. and Methods, 157 (3), 483-6 (1978), PA82-18069.
2205. Maczka, D., Latuszynski, A. and Zuk, W., "Investigation of atom thermoionization in an ion source", Nukleonika, 20 (6), 549-56 (1975), EEA79-22372.
2206. Maczka, D., Zuk, W. and Latuszynski, A., "Utilization of penning effect in ion source for implantation of small doses", Int. Conf. Ion Implantation Semicond., Proc., 1978, p. 536-44, CA90-96418.
2207. Mair, G. L. R., "Emission from liquid metal ion sources", Nucl. Instrum. & Methods, 172 (3), 567-76 (1980), EEA83-51210.
2208. Majundar, N. K., "A pica for charge injection for ion extraction", Nuovo Cimento B, 29B (1), 178-88 (1975), EEA79-6382.
2209. Makov, B. N., "The multiply charged ion source with indirectly heated cathode", IEEE Trans. Nucl. Sci., NS-23 (2), 1035-41 (1976), EEA79-30266.
2210. Mallory, M. L. and Crandall, D. H., "A Penning multiply charged heavy ion source test facility", IEEE Trans. Nucl. Sci., NS-23 (2), 1069-72 (1976), CA84-186231.
2211. Mallory, M. L., Miller, P. S. and Chien, W. S., "A quick change axial cold cathode ion source", IEEE Trans. Nucl. Sci., NS-24 (3), 1593 (1977), EEA80-43908.
2212. Maulding, S. E., "Ion implantation sources and source materials", Report AD-A00431/OSL, Univ. Illinois, Urbana, 1975, 95 pp., PA79-10498.
2213. McKenzie-Wilson, R. B., Prelec, K. and Hruda, R., "Electrode cooling for long pulse high current ion sources", Proceedings of the 8th Symposium on Engineering Problems of Fusion Research, 1979, p. 225-9, EEA83-51221.

2214. Miller, P. A., Mendel, C. W., Swain, D. W. and Goldstein, S. A., "Ion beam generation and focusing", Report SAND-75-5751, Sandia Labs., Albuquerque, N. Mex., 1975, 22 pp., NSA33-5303.
2215. Mourier, G., Boulanger, P., Peyraud, J. and Delcroix, J. L., "Self-accelerating microwave ion and plasma sources", International Conference on Low-Energy Ion beams, 1977, p. 18-23, EEA81-40206.
2216. Mustafa Abd El Baki, M. and Zakhary, S., "The effect of axial magnetic field on improved constricted radial RF source", Nucl. Instrum. & Methods, 154 (3), 509-14 (1978), EEA82-2273.
2217. Naito, M., Murase, K., Nakashima, Y., Nunogaki, M. and Akimune, H., "Operational Characteristics of two-stage discharge ion source", Jpn. J. Appl. Phys., 19 (2), 337-44 (1980), EEA83-26237.
2218. Newton, G. and Unsworth, P. J., "A high-yield ion source for producing beams of hydrogen(+), triatomic hydrogen(+), helium(+), neon(+), argon(+), lithium(+), potassium(+), sodium(+), and cadmium(+)", J. Appl. Phys., 47 (1), 70-3 (1976), CA84-169898.
2219. Noskov, D. A. and Pankovets, N. G., "Ion projector", Instrum. & Exp. Tech., 21 (5), pt. 2, 1361-2 (1978), EEA83-11795.
2220. Okada, C., Kamada, K., Ono, H., Ishizuka, H., Nemoto, N., Kubota, Y. and Miyahara, A., "Optimum anode-cathode spacing in a reflex triode ion source", Jap. J. Appl. Phys., 15 (8), 1615-16 (1976), EEA79-43316.
2221. Okumura, Y., Mizutani, Y. and Ohara, Y., "Experimental study of ion beamlet steering by aperture displacement in two-stage accelerator", Rev. Sci. Instrum., 51 (4), 471-3 (1980), EEA83-37144.
2222. Ollison, W. M., Peters, W. A. and Cross, R. J., Jr., "Regulation of arc or emission current in hot cathode ion sources", Rev. Sci. Instrum., 47, 971 (1976).
2223. Orr, F. D., "Low Perverance ion source bridges low and high intensities in ion implantation", Electrochemical Society Spring Meeting, Extended Abstracts, 1976, p. 532, PA79-91569.
2224. Pasztor, E., "Cathode-lifetime and stability problems in arc-discharge-type heavy ion sources (for ion implantation)", Radiat. Eff., 44 (1-4), 121-8 (1979), PA83-32558.
2225. Popov, V. F., Tserpitskii, B. D. and Raitsyn, A. B., "Investigation of duoplasmatron as a source of argon ions in an ion microanalyzer", Instrum. & Exp. Tech., 20 (3), pt. 2, 814-16 (1977), EEA81-44010.
2226. Prajzner, A. and Zdanowski, J., "Duoplasmatron ion source", Pr. Nauk. Inst. Technol. Elektron. Politech. Wroclaw, 21, 41-8 (1978), CA90-196493.
2227. Purser, K. H., "Heavy ion injection for tandem accelerators", Nucl. Instrum. and Methods, 146 (1), 115-19 (1977), EEA81-12093.

2228. Rasser, B., Pearson, D. I. C. and Remy, M., "Surface ionization source for heavy ions", Rev. Sci. Instrum., 51, 474-7 (1980).
2229. Rochet, D. and Schwerer, A., "Modification of standard ion sources for low current applications", IBM Tech. Disclosure Bull., 20 (4), 1469 (1977), EEA81-25957.
2230. Sakudo, N., Tokiguchi, K., Koike, H. and Kanomata, I., "Microwave ion source", Rev. Sci. Instrum., 48 (7), 762-6 (1977), PA81-327.
2231. Sakudo, N., Tokiguchi, K., Koike, H. and Kanomata, I., "Microwave ion source for high-current implanter", Rev. Sci. Instrum., 49 (7), 940-3 (1978), EEA81-42358.
2232. Schnitzer, R. and Engesser, F. C., "Bakeable hollow cathode duoplasmatron ion source", Rev. Sci. Instrum., 47 (9), 1219-21 (1976), CA85-152590.
2233. Schulte, H., Jacoby, W. and Wolf, B. H., "Development of Penning multiply charged ion sources for the UNILAC", IEEE Trans. Nucl. Sci., NS-23 (2), 1042-8 (1976), CA84-186227.
2234. Schulte, H., Wolf, B. H. and Winter, H., "Ion and electron flow in hot-cathode PIG sources for multiply charged heavy ions", IEEE Trans. Nucl. Sci., NS-23 (2), 1053-60 (1976), CA84-186229.
2235. Sidenius, G., "Ion sources for low energy accelerators", Nucl. Instrum. & Methods, 151 (3), 349-62 (1978), EEA81-35700.
2236. Sugiura, H., "Simple, high current, antimony ion source", Rev. Sci. Instrum., 50 (1), 84-5 (1979), EEA82-14825.
2237. Swaminathan, S. and Venkatasubramanian, V. S., "Ion energy characteristics of a Penning ion source", J. Indian Inst. Sci., 58 (2), 37-46 (1976), EEA80-18431.
2238. Takagi, T., "Multibeam ion source with multiple openings", Patent Germany 2645346, Publ. May 1977, CA87-210266.
2239. Toyoda, K., Kim, P. H., Okabe, Y. and Namba, S., "Ion sources utilizing TEA-CO_2 laser irradiation", Rep. Inst. Phys. & Chem. Res., 53 (3), 70-7 (1977), Japanese, EEA80-43933.
2240. Tsai, C. C., Stirling, W. L. and Ryan, P. M., "Plasma studies on a duoPIGatron ion source", Rev. Sci. Instrum., 48 (6), 651-5 (1977), EEA80-38402.
2241. Williams, N., "A high-current ion source for implantation applications", International Conference on Low-Energy Ion beams, 1977, p. 70-7, EEA81-40212.
2242. Williams, N., "High-current ion source for use on the PR-30 implanter", J. Vac. Sci. & Technol., 15 (3), 1076-9 (1978), EEA81-42350.
2243. Wilson, R. G., "Versatile ion source manifold and rapid ion beam interchange", Proc. Symp. Ion Sources Form. Ion Beams, 2nd, Vol. VII8, 1974, 4 pp., Publ. 1975, CA85-201125.
2244. Winters, H. F., "Hot filament, arc type ion source and method", Patent USA 4103042, Publ. August 1978.
2245. Witteveen, G. J., "A note on ion beam extraction from an RF ion source", Nucl. Instrum. & Methods, 158, 51-6 (1979).

2246. Wittmaack, K. and Wach, W., "Determination of ion source pressure from discharge characteristics", Nucl. Instrum. and Methods, 143 (1), 7-12 (1977), EEA80-30336.
2247. Wokock, H. D., "A controllable ion gun with high brightness", J. Phys. E, 10 (10), 973-6 (1977), EEA81-3339.
2248. Yamanouchi, M., Higashi, Y., Yamaguchi, H. and Okamoto, M., "A new version of sputter ion source utilizing an ionizer of platinum plate", Nucl. Instrum. and Methods, 158 (2-3), 339-45 (1979), EEA82-14822.
2249. Yatsu, K., Nozaki, Y., Hagiwara, S. and Miyoshi, S., "Extraction of an ion beam from a diverter-type plasma source", Rev. Sci. Instrum., 48 (5), 571-2 (1977), EEA80-38401.

B. Large Area

2250. Bickes, R. W., Jr., Bacon, F. M. and O'Hagan, J. B., "Gas discharge ion source. III. Modified Berkeley multifilament ion source", Rev. Sci. Instrum., 49 (11), 1513-17 (1978), EEA82-6051.
2251. Davis, R. C., Jernigan, T. C., Morgan, O. B., Stewart, L. D. and Stirling, W. L., "DuoPIGatron. II. Ion source", Rev. Sci. Instrum., 46 (5), 576-81 (1975), EEA78-34979.
2252. Dembinski, M. and John, P. K., "Efficient ion beam extraction from a flowing plasma", J. Appl. Phys., 50 (10), 6113-16 (1979), PA83-18044.
2253. Dembinski, M., John, P. K. and Ponomarenko, A. G., "High-current ion beam from a moving plasma", Appl. Phys. Lett., 34 (9), 553-5 (1979), PA82-68361.
2254. Ehlers, K. W., Baker, W. R., Berkner, K. H., Cooper, W. S., Kunkel, W. B., Pyle, R. V. and Stearns, J. W., "Large-area plasma sources", Proceedings of the Second Symposium on Ion Sources and Formation of Ion Beams, 1974, Pap. No. I.5.1, NSA33-25610.
2255. Ehlers, K. W. and Leung, K. N., "Characteristics of the Berkeley multicusp ion source", Rev. Sci. Instrum., 50 (11), 1353-61 (1979), PA83-54040.
2256. Goede, A. P. H., Green, T. S. and Singh, B., "Measurement of electron energy distribution in a magnetic multipole ion source", J. Appl. Phys., 51 (4), 1896-9 (1980), PA83-74058.
2257. Haughian, J. M., Baker, W. R., Biagi, L. A., Hopkins, D. B., Owren, H. M. and Paterson, J. A., "Update on the development of 120-keV multi-megawatt neutral beam source", IEEE Trans. Nucl. Sci., NS-24 (3), 1034-6, 8 (1977), PA80-83357.
2258. Kulygin, V. M., Panassenkov, A. A., Semashko, N. N., Boldasov, V. S., Volkov, B. I. and Sveshnikov, A. G., "Formation of intense ion beam with a multiple-slot extractor", Proceedings of the Second Symposium on Ion Sources and Formation of Ion Beams, 1974, Pap. No. II.10.1, NSA33-27198.

2259. Leung, K. N., Collier, R. D., Marshall, L. B., Gallaher, T. N., Ingham, W. H., Kribel, R. E. and Taylor, G. R., "Characteristics of a multidipole ion source", Rev. Sci. Instrum., 49 (3), 321-5 (1978), PA81-55080.
2260. Leung, K. N., Collier, R. D., Taylor, G. R. and Kribel, R. E., "Increasing the ion beam current from a multidipole source", Appl. Phys. Lett., 31 (3), 154-5 (1977), PA80-79489.
2261. Nunogaki, M., Naito, M., Akimune, H. and Suita, T., "Steady-state extraction of multi-ampere ion beams from the BPD source plasma", Proceedings of the Second Symposium on Ion Sources and Formation of Ion Beams, 1974, Pap. No. VI.5.1, NSA33-25680.
2262. Oka, Y. and Kuroda, T., "Effects of a multipole magnetic field on characteristics of a multifilament plasma source", Appl. Phys. Lett., 34 (2), 134-6 (1979), PA82-46462.
2263. Pak, S. and Sites, J. R., "Broad beam ion source operation with four common gases", Rev. Sci. Instrum., 51 (4), 536-9 (1980).
2264. Poulson, P. and Hooper, E. B., Jr., "Large area negative ion source for high voltage neutral beams", Proceedings of the 8th Symposium on Engineering Problems and Fusion Research, II, 1979, p. 676-9, PA83-99379.
2265. Schoenberg, K. F. and Kunkel, W. B., "General plasma characteristics of a Berkeley multifilament ion source", J. Appl. Phys., 50 (7), 4685-91 (1979), PA82-94807.
2266. Stirling, W. L., Ryan, P. M., Tsai, C. C. and Leung, K. N., "Magnetic multipole line-cusp plasma generator for neutral beam injectors", Rev. Sci. Instrum., 50 (1), 102-8 (1979), PA82-19056.
2267. Stirling, W. L., Tsai, C. C. and Ryan, P. M., "15 cm duoPIGatron ion sources", Rev. Sci. Instrum., 48 (5), 533-6 (1977).
2268. Vahrenkamp, R. P. and Seliger, R. L., "A 100-mA low-emittance ion source for ion-beam fusion", IEEE Trans. Nucl. Sci., NS-26 (3), pt. 1, 3101-3 (1979), EEA82-35971.

C. Negative Ion

2269. Alessi, J. G., Laughlin, W., Saladin, J. X., McNamee, W. L. and Gazarik, S. D., "A negative ion sputter source for operation in a Van de Graeff terminal", Nucl. Instrum. & Methods, 147 (2), 305-12 (1977), EEA81-12105.
2270. Alton, G. D., "Recent negative ion source developments", IEEE Trans. Nucl. Sci., NS-26 (1), pt. 2, 1542-7 (1979), PA82-49884.
2271. Alton, G. D. and Blazey, G. C., "Studies associated with the development of a modified University of Aarhus negative ion source", Nucl. Instrum. and Methods, 166 (2), 105-16 (1979), EEA83-11796.

2272. Brand, K., "The reflected beam sputter source", Nucl. Instrum. & Methods, 142 519 (1977).
2273. Brand, K., "Improvement of the reflected beam sputter source", Nucl. Instrum. & Methods, 154 (3), 595-6 (1978), EEA82-2274.
2274. Braun-Elwert, G., Huber, J., Korschinek, G., Kutschera, W., Goldstein, W. and Hershberger, R. L., "The negative-ion test injector of the Munich MP tandem and the Hiconex 834 sputter source", Nucl. Instrum. and Methods, 146 (1), 121-38 (1977), EEA81-12103.
2275. Caskey, G. T., Douglas, R. A., Richards, H. T. and Smith, H. V., Jr., "A simple negative-ion sputter source", Nucl. Instrum. & Methods, 157 (1), 1-7 (1978), EEA82-10267.
2276. Chapman, K. R., "Development of the inverted sputter source", IEEE Trans. Nucl. Sci., NS-23 (2), 1109-12 (1976), EEA79-26283.
2277. Ehlers, K. W. and Leung, K. N., "Multicusp negative ion source", Rev. Sci. Instrum., 51, 721-7 (1980), EEA83-46259.
2278. Kobayashi, M., Peelec, K. and Sluyters, T., "Studies of hollow duoplasmatron as a source of H^- ions", Rev. Sci. Instrum., 47, 1425-34 (1976).
2279. Middleton, R., "A review of sputter negative ion source", IEEE Trans. Nucl. Sci., NS-23, 1098-103 (1975).
2280. Middleton, R., "A survey of negative ions from a cesium sputter source", Nucl. Instrum. & Methods, 144 (3), 373-99 (1977), EEA80-43914.
2281. Middleton, R., "On producing negative ion beams of magnesium and calcium in a cesium sputter source", Nucl. Instrum. and Methods, 141 (2), 373-5 (1977).
2282. Pelletier, J. and Pomot, C., "Intense halogen negative ion source", International Conference on Low-Energy Ion beams, 1977, p. 24-8, EEA81-40207.
2283. Pelletier, J., Pomot, C. and Cocagne, J., "Negative surface ionization: intense halogen-ion source", J. Appl. Phys., 50 (7), 4517-23 (1979).
2284. Prelec, K., "A Penning source for negative hydrogen ions", Nucl. Instrum. & Methods, 144, 359 (1977).
2285. Prelec, K., Shuyters, T. and Grossman, M., "High current negative ion beams", IEEE Trans. Nucl. Sci., NS-24 (3), 1521-3 (1977), EEA80-43902.
2286. Rachidi, I., Monte, J., Pelletier, J., Poinot, C. and Rinchet, F., "Source ionization negative ion source", Appl. Phys. Lett., 28, 292-4 (1976).
2287. Sherwood, J. E. and Zwicker, R. D., "Hollow beam prototype of the universal negative ion source for tandem accelerators", Nucl. Instrum. & Methods, 129, 43-45 (1975).
2288. Vernon Smith, H., Jr., "A modified Aarhus negative ion source", Nucl. Instrum. & Methods, 164, 1 (1979).

D. Numerical Simulation

2289. Asai, K., Yasaka, Y., Inoue, N. and Uchida, T., "Computer simulation of ion beam extraction from free plasma surface", Jpn. J. Appl. Phys., 15, 1343-7 (1976).

2290. Boers, J. E., Freeman, J. R. and Poukey, J. W., "Computer simulation of focused ion beams", Proceedings of the Second Symposium on Ion Sources and Formation of Ion Beams, 1974, Pap. No. II.2.1, NSA33-25614.

2291. Conrad, J. R., "Beamlet steering by aperture displacement in ion sources with large acceleration-deceleration ratio", Rev. Sci. Instrum., 51 (4), 418-24 (1980), PA83-63430.

2292. Cooper, W. S., Halbach, K. and Magyary, S. B., "Computer-aided extractor design", Proceedings of the Second Symposium on Ion Sources and Formation of Ion Beams, 1974, Pap. No. II.1.7, NSA33-25613.

2293. Green, T. S., "Beam formation and space charge neutralization", IEEE Trans. Nucl. Sci., NS-23 (2), 918-28 (1976), CA85-11255.

2294. Jaeger, E. F. and Whitson, J. C., "Numerical simulation for axially symmetric beamlets in the duoPIGatron ion source", Report ORNL-TM-4990, Oak Ridge National Lab., Tenn., 1975, 47 pp., NSA32-25745.

2295. Kim, J., Whealton, J. H. and Schilling, G., "A study of two-stage ion-beam optics", J. Appl. Phys., 49 (2), 517-24 (1978).

2296. Klenov, G. I., Larionov, V. P. and Lyzin, V. V., "Numerical solution to the problem of finding the plasma boundary in ionic sources", Tr. Vses. Soveshch. Uskorit. Zaryazhennykh Chastits, 4th, Vol. 1, 1974, p. 302-5, Publ. 1975, Russian, CA85-150516.

2297. Ohara, Y., "Numerical simulation for design of a two-stage acceleration system in a megawatt power ion source", J. Appl. Phys., 49 (9), 4711-17 (1978).

2298. Pagel, L. and Scheit, U., "Numerical determination of potential fields, ionic paths and plasma boundaries in the extraction system of some ion sources", Wiss. Z., Tech. Hochsch., Karl-Marx Stadt, 17 (3), 349-54 (1975), German, CA84-98465.

2299. Paul, A. C., "Transport: an ion optic program", Report LBL-2697, California Univ., Berkeley, Lawrence Berkeley Lab., 1975, NSA32-23021.

2300. Semashko, N. N., Panasenkov, A. A. and Kulygin, V. M., "Some aspects of stationary ion source construction", Proceedings of the 8th Symposium on Engineering Problems of Fusion Research, 1979, p. 221-4, EEA83-51220.

2301. Whealton, J. H., "Linear optics theory of ion beamlet steering", Rev. Sci. Instrum., 48 (11), 1428-9 (1977), EEA81-10900.

2302. Whealton, J. H., "Effect of beamlet-beamlet interaction on ion optics of multiaperture sources", Appl. Phys. Lett., 32 (6), 353-5 (1978).

2303. Whealton, J. H., "Effect of electrode shielding on beamlet-beamlet interaction in multiaperture sources", Appl. Phys. Lett., 33 (8), 697-8 (1978).
2304. Whealton, J. H., Grisham, L. R., Tsai, C. C. and Stirling, W. L., "Effect of preacceleration voltage upon ion-beam divergence", J. Appl. Phys., 49 (6), 3091-101 (1978).
2305. Whealton, J. H., Jaeger, E. F. and Whitson, J. C., "Optics of single-stage accelerated ion beams extracted from a plasma", Rev. Sci. Instrum., 48 (7), 829-31 (1977).
2306. Whealton, J. H., McGaffy, R. W. and Jaeger, E. F., "Comparison of ion optics for transverse slot and cylinders", Appl. Phys. Lett., 36 (1), 91-3 (1980).
2307. Whealton, J. H., Tsai, C. C., Dagenhart, W. K., et al., "Effect of preacceleration on intense ion-beam transmission efficiency", Appl. Phys. Lett., 33 (4), 278-9 (1978).
2308. Whealton, J. H. and Whitson, J. C., "Extraction of aberrationless high-perveance ion beams from a quiscent equilibrium plasma", J. Appl. Phys., 50 (6), 3964-6 (1979).

3. Beam Transport/Focusing/Scanning

2309. Armitage, S. A. and Eastham, D. A., "The optical properties of a double focusing bending magnet measured with a thin alpha source", Nucl. Instrum. and Methods, 151 (1-2), 61-3 (1978), EEA81-35692.
2310. Baier, H., Bayer, T., Hinkel, H., Kraus, G. and Seybold, D., "Ion beam scanning in a circular pattern", IBM Tech. Disclosure Bull., 23 (4), 1481-2 (1980).
2311. Beanland, B. G. and Freeman, J. H., "Ion beam studies. IIIa: The modelling of electrostatic mirrors for the manipulation and focusing of heavy ions", Nucl. Instum. & Methods, 136 (3), 409-19 (1976), EEA79-49141.
2312. Bernheim, M., "Ion imaging for ion implantation and for micromachining", Radiat. Eff., 44 (1-4), 11-18 (1979), French, EEA83-14296.
2313. Booth, R. and Lefevre, H. W., "Space charge lens for high current ion beams", Nucl. Instrum. & Methods, 151 (1-2), 143-7 (1978), EEA81-35694.
2314. Brown, K. L., Rothacker, F., Carey, D. C. and Iselin, C., "The effect of beam line magnet misalignments", Nucl. Instrum. and Methods, 141 (3), 393-9 (1977), PA80-30688.
2315. Dunkel, H., Rohrer, L. and Korka, S. J., "Effect of misalignments of beam optical components", Nucl. Instrum. and Methods, 146 (1), 97-9 (1977), EEA81-12092.
2316. Ehlers, K. W. and Leung, K. N., "Low energy ion beam transport by permanent magnets", Rev. Sci. Instrum., 51 (4), 500-3 (1980), EEA83-37145.

2317. Elayi, A. G., "Importance of the focusing reduction in an extended fringing field", Nucl. Instrum. & Methods, 135 (2), 285-7 (1976), French, PA79-78407.
2318. Forrester, A. T., Perel, J. and Mahoney, J., "A critique of the throughput limit in mass spectrographs", J. Appl. Phys., 49 (6), 3046-8 (1978), PA81-65995.
2319. Freeman, J. H., Beanland, D. G., Chivers, D. J. and Gard, G. A., "Ion beam studies. VII. An electrostatic lens for the acceleration and deceleration of high intensity ion beams", Nucl. Instrum. & Methods, 155 (1-2), 29-37 (1978), EEA81-50313.
2320. Freeman, J. H., Temple, W., Beanland, D. and Gard, G. A., "Ion beam studies. Pt. I: The retardation of ion beams to very low energies in an implantation accelerator", Report AERE-R 8287, AERE, U. K. At. Energy Auth., Harwell/Oxfordshire, England, 1976, 14 pp., CA85-55459.
2321. Gabovich, M. D., Gasanov, I. S. and Protsenko, I. M., "Forming and focusing of an intensive ion beam in a system with a plasma lens", Pis'ma V Zh. Tech. Fiz., 3 (21), 1153-5 (1977), Russian, EEA81-16834.
2322. Gardner, W. L., Kim, J., Menon, M. M. and Whealton, J. H., "Ion beamlet steering by aperture displacement for a tetrode accelerating structure", Rev. Sci. Instrum., 49 (8), 1214-15 (1978), EEA81-47522.
2323. Green, T. S., "Beam formation and space charge neutralization", IEEE Trans. Nucl. Sci. NS-23, 918-28 (1975).
2324. Gross, E. E., "Calibration of an analysing magnet using an α-source and heavy-ion beams", Nucl. Instrum. Methods, 135 (2), 401-2 (1976), CA85-84409.
2325. Gupta, S. K., Bhatia, M. S. and Kerekatte, S. S., "Measurement of the apparent non-linearity of analyzing magnets", Nucl. Instrum. & Methods, 131 (3), 403-6 (1975), EEA79-26271.
2326. Heck, D., "Ion beam optics of misaligned beam transport elements", Nucl. Instrum and Methods, 143 (3), 423-7 (1977), PA80-65564.
2327. Holmes, A. J. T., "Theoretical and experimental study of space charge in intense ion beams", Phys. Rev. A, 19 (1), 389-407 (1979), PA82-41325.
2328. Johnson, C. H., "A ring lens for focusing ion beams to uniform densities", Nucl. Instrum. Methods, 127 (2), 163-71 (1975), NSA33-9970.
2329. Keller, J. H., "Beam scanning-electrostatic", Radiat. Eff., 44 (1-4), 71-80 (1979), EEA83-14299.
2330. Larson, J. D., "Resolving beam transport problems in electrostatic accelerators", Rev. Phys. Appl., 12 (10), 1551-61 (1977), EEA81-21306.
2331. Lefevre, H. W., "Progress in space charge lens development", IEEE Trans. Nucl. Sci., NS-26 (3), pt. 1, 3115-17 (1979), EEA82-35873.

2332. Liebl, H., "A modified cylindrical condenser as electrostatic prism", Int. J. Mass Spectrom. & Ion Phys., 22 (1-2), 203-7 (1976), EEA80-3253.
2333. Pavinskii, P. P., "Calculation of focusing magnetic field for a flat beam of charged particles", Vestn. Leningr. Univ. Fiz. & Khim., no. 1, 59-66 (1975), Russian, EEA78-39278.
2334. Pearce, R. M., "Reduction of multiple scattering displacement by a magnetic field parallel to the beam", Nucl. Instrum. and Methods, 164 (1), 11-14 (1979), EEA82-44345.
2335. Poukey, J. W. and Humphries, S., Jr., "Fast neutralization of ion beams in the presence of transverse magnetic fields", Appl. Phys. Lett., 33 (2), 122-4 (1978), EEA81-43989.
2336. Rose, P. H., "Application of mechanical scanning to ion implantation", Radiat. Eff., 44 (1-4), 137-44 (1979), EEA83-19001.
2337. Sakudo, N. and Hayashi, T., "Quadrupole electrodes with flat faces", Rev. Sci. Instrum., 46 (8), 1060-2 (1975), EEA79-10729.
2338. Samsonov, G. A., Sysoev, A. A. and Kuznetsov, V. N., "The effect of the slope of the field boundaries of an electrostatic analyzer on its ion-optic characteristics", Instrum. & Exp. Tech., 18 (2), 376-8 (1975), EEA79-10744.
2339. Segalov, Z., "Ion optics of displaced quadrupole lens charge selector", Nucl. Instrum. & Methods, 130 (2), 607-9 (1975), EEA79-18241.
2340. Segalov, Z. and Skurnik, E., "Ion optics in a 14 UD Pelletron terminal", Rev. Phys. Appl., 12 (10), 1563-6 (1977), EEA81-21307.
2341. Steinhoff, J., "Optimization of heavy ion focussing systems with large numbers of magnets", IEEE Trans. Nucl. Sci., NS-26 (3), pt. 2, 3688-91 (1979), EEA82-35909.
2342. Stewart, L. D., Kim, J. and Matsuda, S., "Beam focusing by aperture displacement in multiampere ion sources", Rev. Sci. Instrum., 46 (9), 1193-6 (1975), PA79-16466.
2343. Taya, S., "An experimental study on the stigmatic focusing condition of a magnetic sector mass analyzer", Nucl. Instrum. and Methods, 152 (2-3), 399-405 (1978), EEA81-44062.
2344. Taya, S., Tokiguchi, K., Kanomata, I. and Matsuda, H., "A triple focusing mass separator", Nucl. Instrum. and Methods, 150 (2), 165-71 (1978), PA81-45748.
2345. Taya, S., Tsuyama, H., Kanomata, I., Noda, T. and Matsuda, H., "A stigmatic, second-order, double-focusing mass spectrometer", Int. J. Mass Spectrom. and Ion Phys., 26 (1), 77-90 (1978), PA81-21256.
2346. Terzic, I. and Ciric, D., "The double cylindrical electrostatic sector as an ion energy analyzer", Nucl. Instrum. & Methods, 166 (3), 419-23 (1979), EEA83-16131.

2347. Thompson, E., "Ion optics of a single stage post-acceleration system", Proceedings of the Second Symposium on Ion Sources and Formation of Ion Beams, 1974, Pap. No. II.7.1, NSA33-25674.

2348. Tokiguchi, K., Sakudo, N., Taya, S., Koike, H. and Kanomata, I., "Ion beam image transformation using magnetic quadrupole lenses", Nucl. Instrum. & Methods, 151 (1-2), 47-52 (1978), EEA81-35691.

2349. Tokiguchi, K., Taya, S., Sakudo, N., Koike, H. and Kanomata, I., "Influence of space charge neutralization on mass separator transmission", Nucl. Instrum. and Methods, 154 (1), 13-17 (1978), PA81-82073.

2350. Vlasov, A. D., "Static electric focusing in low-energy linear accelerators", Sov. Phys. -Tech. Phys., 20 (3), 420-21 (1975), NSA33-23911.

2351. Winnard, J. R., "Ion-beam scanning by magnetic means", IBM Tech. Disclosure Bull., 19 (9), 3411-12 (1977), EEA80-36222.

2352. Wittmaack, K., "Low energy ion beam transport through apertures", Nucl. Instrum. and Methods, 143 (1), 1-6 (1977), EEA80-30332.

2353. Zubkov, I. P., "Structure of the ion beam from a four-lens accelerator", Sov. Phys.-Tech. Phys., 22 (8), 1006-10 (1977), EEA81-40198.

4. Beam Target Systems/Profiling/Dosimetry

2354. Ames, L. L., "Rapid measurement of emittance and brightness", Nucl. Instrum. and Methods, 151 (3), 363-9 (1978), EEA81-35695.

2355. Anno, R., Lefol, A., Milleret, G. and Perret, R., "A low-matter beam profile secondary emission monitor", Nucl. Instrum. & Methods, 152 (2-3), 395-8 (1978), EEA81-44005.

2356. Antipov, Y. M., Bushnin, Y. B., Dunaitsev, A. F., Rybakov, V. G., Senko, V. A., Sytin, A. N., Soldatov, M. M. and Kholkin, A. N., "An installation for measuring the profile of a particle beam", Instrum. & Exp. Tech., 18 (4), 1027-8 (1975), EEA79-26307.

2357. Badalec, R. and Runge, H., "A new method to monitor the beam profile in high-current ion implantation systems", J. Phys. E, 12 (12), 1146-7 (1979), PA83-13182.

2358. Baier, H., Hinkel, H. and Kraus, G., "Measuring target vibrations during ion implantation", IBM Tech. Disclosure Bull., 22 (4), 1523-4 (1979), EEA83-14289.

2359. Bassett, R., "Analysis of beams from a dynamitron accelerator with a view to modification for high precision electron and positive ion operation", Rev. Phys. Appl., 12 (10), 1418 (1977), EEA81-21305.

2360. Bayer, E. H., Kranik, J. R. and Mueller, W. F., "Mechanism-free target disc for centrifugal holding of wafers during ion implantation", IBM Tech. Disclosure Bull., 21 (11), 4501-2 (1979), EEA83-736.
2361. Bayer, E. H., Kranik, J. R. and Mueller, W. F., "Heat transfer pillow for use in high current ion implantation", IBM Tech. Disclosure Bull., 22 (9), 4073 (1980).
2362. Beesarabskii, Y. G., Ovsyannikov, S. B. and Sukhoveev, S. P., "Study of profile of an ion beam by means of a titanium tritide collector", Instrum. & Exp. Tech., 20 (3), pt. 2, 880-2 (1977), EEA81-44002.
2363. Bjarngard, B. E., Chen, G. T. Y. and Maddox, B. J., "High resolution beam profile measurements using phosphorescence", IEEE Trans. Nucl. Sci., NS-22 (3), 1558-60 (1975), NSA32-28860.
2364. Brendle, M. and Steidle, G., "Computer-controlled target-positioning system", Nucl. Instrum. Methods, 126 (1), 93-98 (1975), NSA32-24747.
2365. Broadhurst, J. H., "A high stability vibrating wire profile monitor", Nucl. Instrum. & Methods, 172 (3), 459-72 (1980), EEA83-51206.
2366. Bruel, M., Berthet, B., Floccari, M. and Michaud, J. F., "Target heating during ion implantation and related problems", Radiat. Eff., 44 (1-4), 173-80 (1979), EEA83-14304.
2367. Cairns, J. A., Lurio, A. and Ziegler, J. F., "Method and apparatus for monitoring the dose of ion implanted into a target by counting emitted X-rays", Patent USA 4110625, Publ. August 1978.
2368. EerNisse, E. P., Peterson, G. D. and Schueler, D. G., "Ion beam profile monitor (for accelerators)", Rev. Sci. Instrum., 46 (3), 266-8 (1975), EEA78-26241.
2369. Fine, J. and Gorden, R., Jr., "Visual ion-beam images produced by electron and ion-beam interaction on surfaces", J. Appl. Phys., 49 (3), pt. 1, 1236-40 (1978), PA81-58091.
2370. Fine, J. and Gorden, R., Jr., "Ion and electron beam interaction on surfaces - a detection mechanism for obtaining visual ion beam images", Nucl. Instrum. and Methods, 149 (1-3), 679-83 (1978), EEA82-2280.
2371. Forneris, J. L., Hicks, W. W., Keller, J. H., McKenna, C. M. and Seirmarco, J. A., "Ion implantation apparatus for controlling the surface potential of a target surface", Patent USA 4135097, Publ. January 1979.
2372. Fortino, A. G. and Kingsland, S. F., "Ion implantation beam monitor", IBM Tech. Disclosure Bull., 21 (6), 2254 (1978), EEA82-22080.
2373. Freeman, J. H. and Booker, D. V., "Ion beam studies. V. The computer simulation of composite ion implantation profiles", Nucl. Instrum. & Methods, 144 (2), 175-81 (1977), EEA80-41392.

2374. Geipel, H. and Warley, S. D., "In-line ion-beam monitor", IBM Tech. Disclosure Bull., 20 (5), 1712-13 (1977), EEA81-33266.

2375. Gruen, D. M., Krauss, A. R. and Siskind, B., "Method and means of directing an ion beam onto an insulating surface for ion implantation or sputtering", Report PAT-APPL-812 936-SL, Department of Energy, Washington, DC, 1977, 10 pp.

2376. Hemment, P. L. F., "Absolute dosimetry of ion implanted impurities using a calorimetric method", Vacuum, 27 (10-11), 611-16 (1977), EEA81-14662.

2377. Hemment, P. L. F., Stephens, K. G., Wilson, I. H. and Moruzzi, J. L., "Calorimetric measurements of the neutral beam present in a 500 keV ion implanter", International Conference on Low Energy Ion Beams, 1978, p. 117-24, PA82-30605.

2378. Higgins, E. F. and Griffin, J. E., "High-speed beam position electronics for accelerator control and diagnostics", IEEE Trans. Nucl. Sci., NS-22 (3), 1574-77 (1975), NSA32-28863.

2379. Hlavka, J., "Ion implantation. Measurement of ion pressure", Phys. Lett. A, 45 (5), 407-8 (1973), CA80-42156.

2380. Hodgart, M. S., Andrews, A. and Haining, R. W., "Remote ion beam position measurement from random beam modulation", International Conference on Low Energy Ion Beams, 1977, p. 125-30, EEA82-19774.

2381. Hoff, P., Jacobsson, L., Johansson, B., Aagaard, P., Rudstam, G. and Zwicky, H.-L., "Production of molecular ions in the integrated target/ion-source of an isotope-separation-on-line system", Nucl. Instrum. & Methods, 172 (3), 413-18 (1980), EEA83-51209.

2382. Jamba, D. M., "Some aspects of dose measurement for accurate ion implantation", Report NBS-SP-410-39, Washington, DC, 1977, 36 pp., EEA81-1207.

2383. Jamba, D. M., "Secondary particle collection in ion implantation dose measurement", Rev. Sci. Instrum., 49 (5), 634-8 (1978), PA81-67731.

2384. Ko, W. C. and Sawatzky, E., "High current density ion beam measurement techniques (ion implantation)", Electrochemical Society Spring Meeting, Extended Abstracts, 1976, p. 457-8, EEA79-46100.

2385. Kranik, J. R., "Design of an end station for a high current ion implantation system", Radiat. Eff., 44 (1-4), 81-92 (1979), EEA83-14300.

2386. Kranik, J. R. and Mueller, W. F., "Apparatus for automatically balancing wafers during rotational ion implantation", IBM Tech. Disclosure Bull., 20 (12), 5180-1 (1978).

2387. Larsen, R. A., "Ion-implant dose monitor", IBM Tech. Disclosure Bull., 20 (8), 3019 (1978), EEA82-4298.

2388. Lasinger, M., Paul, H. and Semrad, D., "Beam profiling by back-scattering from thin wires", Nucl. Instrum. & Methods, 161 (2), 199-201 (1979), EEA82-32081.

2389. Matteson, S., Tonn, D. G. and Nicolet, M. A., "Diagnostic test for ion implantation dosimetry", J. Vac. Sci. and Technol., 16 (3), 882-3 (1979), EEA82-47081.
2390. McKenna, C. M., "High current dosimetry techniques for ion implantation", Radiat. Eff., 44 (1-4), 93-110 (1979), EEA83-14301.
2391. McKenna, C. M. and Mueller, W. F., "Ion implantation apparatus with a cooled structure controlling the surface potential of a target surface", Patent USA 4118630, Publ. October 1978.
2392. Miller, G. H. and Lockwood, G. J., "Calorimetric determination of beam energy", IEEE Trans. Nucl. Sci., NS-22 (3), 1072-75 (1975), NSA32-28814.
2393. Myers, S. M., "Ion-beam-induced migration and its effect on concentration profiles", Nucl. Instrum. & Methods, 168 (1-3), 265-74 (1980), PA83-59723.
2394. Natsuaki, N., Ohyu, K. and Tokuyama, T., "Spatial dose uniformity monitor for electrically scanned ion beam", Rev. Sci. Instrum., 49 (9), 1300-4 (1978), EEA81-51558.
2395. Nishi, H., Inada, T., Sakurai, T., Kaneda, T., Hisatsugu, T. and Furuya, T., "Uniform doping of channeled-ion implantation", J. Appl. Phys., 49 (2), 608-13 (1978), EEA81-28415.
2396. Peterson, G. D. and EerNisse, E. P., "Combined ion-implantation-analysis sample chamber", Rev. Sci. Instrum., 47, 1153-6 (1976).
2397. Riemann, R., "Method for the noninterceptive measurement of an ion beam", Nucl. Instrum. & Methods, 137 (2), 385-6 (1976), German, EEA80-1629.
2398. Rohrer, L. and Schnitter, H., "A light link coupled current monitor", Rev. Phys. Appl., 12 (10), 1415-16 (1977), EEA81-21303.
2399. Rupprecht, H. S. and Schwenker, R. O., "Method of high current ion implantation and charge reduction by simultaneous kerf implant", Patent USA 4076558, Publ. February 1978.
2400. Ryding, G., "Improvements in or relating to target chambers", Patent UK 1389294/B/, Publ. April 1975, NSA33-1487.
2401. Sakudo, N., Kanomata, I., Tsuyama, H. and Mitani, E., "Display device for ion beam profile", Rev. Sci. Instrum., 47, 1288-92 (1976).
2402. Satok, K., "New beam position monitor by detecting wall currents", Rev. Sci. Instrum., 50 (4), 450-3 (1979), EEA82-32082.
2403. Stufflebeam, J. H., Jennings, W. C. and Hickok, R. L., "Ion beam probe diagnostic system for the LITE experiment", Report COO-2444-1, Plasma Dyn. Lab., Rennselaer Polytech. Inst., Troy, N. Y., 1974, 9 pp., CA85-185398.
2404. Takeda, Y., Toda, S. and Shinozuka, T., "Heating accelerator targets by ion beam and its thermal characteristics", Technol. Rep. Tohoku Univ., 44 (2), 471-84 (1979), Japanese, PA83-99374.

2405. Tasic, S. and Petrinovic, M., "System for position and intensity profile measurement of charged particles beam in the ion accelerator", Elektrotehnika, Zagreb, 20 (1), 32-6 (1977), Croation, EEA80-38404.
2406. Turkenburg, W. C., de Haas, E., Neuteboom, A. F., Ladru, J. and Kersten, H. H., "Universal target manipulator for use in ultra-high vacuum", Nucl. Instrum. Methods, 126 (2), 241-45 (1975), NSA32-24744.
2407. Vadehra, S. K. and Wajda, W. W., "Ion-beam-size analyzer", IBM Tech. Disclosure Bull., 20 (10), 3886-8 (1978), EEA82-6044.
2408. Vader, R. J. and Dermois, O. C., "A stepping motor driven ion beam scanner", Nucl. Instrum. & Methods, 144 (3), 423-7 (1977), EEA80-43896.
2409. Varma, M. N., "A simple heavy-ion beam monitor", Nucl. Instrum. & Methods, 133 (2), 205-7 (1976), EEA79-30239.
2410. Weber, G., Robaye, G. and Quaglia, L., "A simple method to test the homogeneity of an accelerated charged particle beam", Nucl. Instrum. & Methods, 148 (1), 13-16 (1978), EEA81-16833.
2411. Wei, C. -Y. and Seidman, D. N., "Novel Faraday cup for the simultaneous observation and measurement of ion-beam currents", Rev. Sci. Instrum., 48 (12), 1617-20 (1977).
2412. Weisberg, H., "Improved measurements of slow-extracted beam emittance", Nucl. Instrum. & Methods, 169 (1), 11-17 (1980), EEA83-21273.
2413. Wells, G. F., "The double transmitting Faraday cup: an ion optical control device", Nucl. Instrum. & Methods, 153 (2-3), 339-41 (1978), EEA81-47517.
2414. Zinevich, N. I. and Cherepanov, V. P., "Use of a dielectric in charged-particle beam-position detectors", Instrum. & Exp. Tech., 21 (2), pt. 1, 296-8 (1978), EEA82-14811.

5. Micro Ion Beams

2415. Augustyniak, W. M., Betteridge, D. and Brown, W. L., "A miniature electrostatic lens for forming MeV millibeams", Nucl. Instrum. and Methods, 149 (1-3), 669-73 (1978), EEA82-1320.
2416. Bernheim, M., "Ion imaging for ion implantation and for micro-machining", Radiat. Eff., 44 (1-4), 11-18 (1979), French, EEA83-14296.
2417. Bonani, G., Suter, M., Jung, H., Stoller, C. and Wolfli, W., "A digitally controlled scanning microprobe for protons and heavy ions", Nucl. Instrum. and Methods, 157 (1), 55-63 (1978), EEA82-10262.
2418. Brune, D., Lindh, U. and Lorenzen, J., "A proton semi-microbeam device for surface analysis", Nucl. Instrum. and Methods, 142 (1-2), 51-4 (1977), PA80-48720.

2419. Chan, T. N., Dousson, S., Geller, R., Jaquot, B. and Lieuvin, M., "Some radial emittance measurements of the multiply charged heavy ion source micromafios", Nucl. Instrum. & Methods, 174 (1-2), 151-6 (1980), EEA83-55605.

2420. Clampitt, R. and Jefferies, D. K., "Molten metal field ion sources", International Conference on Low-Energy Ion Beams, 1977, p. 12-17, EEA81-40205.

2421. Clampitt, R. and Jefferies, D. K., "Miniature ion sources for analytical instruments", Nucl. Instrum. and Methods, 149 (1-3), 739-42 (1978), PA82-3266.

2422. Gomer, R., "On the mechanism of liquid metal electron and ion sources", Appl. Phys., 19 (4), 365-75 (1979), PA82-85845.

2423. Hung, R. Y., McCorkie, R. A. and Ziegler, J. F., "Thin-film field ionization source", IBM Tech. Disclosure Bull., 22 (7), 2926-7 (1979), EEA83-26236.

2424. Keohn, V. E. and Ringo, G. R., "Ion source of high brightness using liquid metal", Appl. Phys. Lett., 27, 479-81 (1975).

2425. Mair, G. L. R. and Von Engel, A., "Gallium-field-ion emission from liquid point anodes", J. Appl. Phys., 50 (9), 5592-5 (1979), PA83-2928.

2426. Nobiling, R., Civelekoglu, Y., Povh, B., Schwalm, D. and Traxel, K., "Collimation of ion beams to micrometer dimensions", Nucl. Instrum. Methods, 130 (2), 325-34 (1975), CA84-113173.

2427. Nobiling, R., Traxel, K., Bosch, F., Civelekoglu, Y., Martin, B., Povh, B. and Schwalm, D., "Focussing of proton beams to micrometer dimensions", Nucl. Instrum. and Methods, 142 (1-2), 49-50 (1977), PA80-46444.

2428. Orloff, J. H. and Swanson, L. W., "Study of a field-ionization source for microprobe applications", J. Vac. Sci. & Technol., 12, 1209-13 (1975).

2429. Orloff, J. and Swanson, L. W., "Fine-focus ion beams with field ionization", J. Vac. Sci. and Technol., 15 (3), 845-8 (1978), EEA81-42210.

2430. Ringo, G. R. and Krohn, V. E., "Possible applications of a high brightness gallium source to ion microprobes", Nucl. Instrum. and Methods, 149 (1-3), 735-7 (1978), PA82-2896.

2431. Sakurai, T., Culbertson, R. J. and Robertson, G. H., "Field-ion microscopy of liquid-metal gallium", Appl. Phys. Lett., 34 (1), 11-13 (1979), PA82-43106.

2432. Seliger, R. L. and Ward, J. W., "Fabrication of lateral doping profiles by a computer-controlled focused ion beam", J. Vac. Sci. & Technol., 12 (6), 1378-81 (1976), EEA79-20321.

2433. Seliger, R. L., Ward, J. W., Wang, V. and Kubena, R. L., "A high-intensity scanning ion probe with submicrometer spot size", Appl. Phys. Lett., 34 (5), 310-12 (1979), PA82-54225.

2434. Swanson, L. W., Schwind, G. A. and Bell, A. E., "Emission characteristics of a liquid gallium ion source", Scanning Electron Microsc., pt. 1, 45-51 (1979), EEA83-28904.

2435. Swanson, L. W., Schwind, G. A., Bell, A. E. and Brady, J. E., "Emission characteristics of gallium and bismuth liquid metal field ion sources", J. Vac. Sci. and Technol., 16 (6), 1864-7 (1979), PA83-74059.

6. Miscellaneous

2436. Akimov, V. M., Volkov, A. I. and Rusin, L. Y., "Power-supply block for ion source with electron bombardment", Instrum. & Exp. Tech., 19 (5), pt. 2, 1586 (1976), EEA81-7844.
2437. Almazov, A. V., Opasin, V. V. and Myntsov, F. F., "System for stabilizing the voltage of electrostatic accelerators", Instrum. & Exp. Tech., 18 (3), 707-10 (1975), EEA79-14482.
2438. Bailey, E. D., "State registration and inspection programs for particle accelerators", Proceedings of the 4th Conference on the Scientific and Industrial Applications of Small Accelerators, 1976, p. 363-9, EEA80-43833.
2439. Bayer, E. H., Frederick, J. J., Jung, G. and Popp, G., "Insulating shadow groove [ion implantation tool]", IBM Tech. Disclosure Bull., 22 (6), 2276-7 (1979), EEA83-23909.
2440. Bayer, E. H., Kranik, J. R. and Mueller, W. F., "Heat transfer pillow for use in high current ion implantation", IBM Tech. Disclosure Bull., 22 (9), 4073 (1980).
2441. Bobbitt, J., "Applications of microprocessors in accelerators", IEEE Trans. Nucl. Sci., NS-22 (3), 1065-8 (1975), EEA78-43645.
2442. Brand, K., "A tanked tube for open air accelerators", Nucl. Instrum. & Methods, 172 (3), 419-21 (1980), EEA83-51205.
2443. Brautti, G., Clauser, T., Raino, A. and Stagno, V., "The delay-line voltage multiplier", Rev. Phys. Appl., 12 (10), 1585 (1977), EEA81-21292.
2444. Briska, M. and Schmitt, A., "Determining the sputter contamination of an ion implantation system", IBM Tech. Disclosure Bull., 22 (5), 1931 (1979), EEA83-18992.
2445. Chu, W., "Very low dose ion implantation by scattering method", IBM Tech. Disclosure Bull., 21 (8), 3188-9 (1979), EEA82-38247.
2446. Croset, M., Dieumegard, D. and Grouille, A., "Energy resolution stability and calibration of a 400 kV implantation-type accelerator studied using (p,γ) nuclear reactions", International Conference on Low Energy Ion Beams, 1977, p. 109-16, Publ. 1978, EEA82-19773.
2447. Crowley-Milling, M. C., "Computer control applied to accelerators", Meeting on Technology Arising from High-Energy Physics, CERN, Proceedings, 1974, p. 120-8, NSA32-9640.
2448. Drummond, C. W. and Bourianoff, G. I., "Reducing computer costs", Ind. Res./Dev., 22 (5), 132-7 (1980), EEA83-51201.

2449. Eastham, D. A. and Joy, T., "The accelerator tube development programme at Daresbury laboratory", Rev. Phys. Appl., 12 (10), 1503-5 (1977), EEA81-21284.
2450. Gardner, J. A., Lever, R. F. and Michel, A. E., "Ion implant heat transfer", IBM Tech. Disclosure Bull., 21 (11), 4526 (1979), EEA83-737.
2451. Goodwin, R. W., Kocanda, R. F. and Shea, M. F., "A microprocessor-based preaccelerator control system", 1976 Proton Linear Accelerator Conference, p. 264-8, EEA80-18385.
2452. Graham, C. L., "Accelerator safety program at the Lawrence Livermore Laboratory", Proceedings of the 4th Conference on the Scientific and Industrial Applications of Small Accelerators, 1976, p. 370-2, EEA80-43834.
2453. Guernet, G. and Lefeuvre, G., "Method and a device for implantation of particles into a substrate", Patent USA 4035655, Publ. July 1977.
2454. Haas, E. W., Glawischnig, H., Lichti, G. and Bleier, A., "Activation analytical investigation of contamination and cross-contamination in ion implantation", J. Electron. Mater., 7 (4), 525-33 (1978), EEA81-45938.
2455. Hida, H., Ohgaku, T., Yoshikuni, A. and Ohtsubo, A., "A stabilization method of ion beam current", Jpn. J. Appl. Phys., 16 (8), 1503-4 (1977), EEA80-43894.
2456. Hinkel, H., Kaus, G., Kempf, J. and Schmitt, A., "Determination of target contaminations by ion implantation", IBM Tech. Disclosure Bull., 21 (11), 4573-4 (1979), EEA83-739.
2457. Hyder, H. R. M., "Accelerator tubes", Rev. Phys. Appl., 12 (10), 1493-1501 (1977), EEA81-21283.
2458. Ilyushkin, V. A., Noskov, D. A., Shangin, A. S. and Shibaev, Y. A., "Multichannel ion-beam device", Instrum. & Exp. Tech., 20 (3), pt. 2, 811-13 (1977), EEA81-44009.
2459. Mallory, K. B., "Applications of microprocessors in upgrading of accelerator controls", IEEE Trans. Nucl. Sci., NS-24 (3), 1677-80 (1977), EEA80-43759.
2460. Michail, M. S., Woodard, O. C. and Yourke, H. S., "Method and apparatus for positioning a beam of charged particles", Patent USA 3900736, Publ. August 1975, NSA33-1478.
2461. Nickerson, E. B., "Safety aspects associated with industrial use of particle accelerators", Proceedings of the 4th Conference on the Scientific and Industrial Applications of Small Accelerators, 1976, p. 342-7, EEA80-43832.
2462. Okamura, H., Kunibe, T. and Kuroda, T., "Acceleration power supply with fast interruption switching circuit for high current ion source", Jap. J. Appl. Phys., 15 (7), 1413-14 (1976), EEA79-39370.
2463. Templeton, L. F., "Vacuum lock design: the key to continuous ion implantation", Solid State Technol., 18 (12), 46-8 (1975), EEA79-8787.

2464. Templeton, L. F. and Gault, R. B., "Apparatus and method for automatic ion implantation of semiconductors", Patent German 2556456, Publ. August 1976, CA85-115653.
2465. Thompson, C. C., Jr. and Malone, H. F., "Filament supply circuit for particle accelerator", Patent USA 3908160, Publ. September 1975, EEA79-22394.
2466. Tsai, M. Y., Streetman, B. G., Blattner, R. J. and Evans, C. A., Jr., "Study of surface contamination produced during high dose ion implantation", J. Electrochem. Soc., 126 (1), 98-102 (1979), PA82-70374.
2467. Ziegler, N., "Improved Van de Graaff belt charge regulator", IEEE Trans. Nucl. Sci., NS-22 (3), 1730-31 (1975), NSA32-28869.

AUTHOR INDEX

SUBJECT INDEX

In this book, references are numbered starting with "1". Entries are given in this index by their number. When the entry is the heading of a section, the page is given underlined.

MIX
Papier aus verantwortungsvollen Quellen
Paper from responsible sources
FSC® C105338

If you have any concerns about our products,
you can contact us on
ProductSafety@springernature.com

In case Publisher is established outside the EU,
the EU authorized representative is:
Springer Nature Customer Service Center GmbH
Europaplatz 3, 69115 Heidelberg, Germany

Printed by Libri Plureos GmbH
in Hamburg, Germany